Principles of AI Digital Twins

Strategy, Governance, and Artificial Intelligence Integration between Real and Virtual Worlds Playbook

How Data, AI, and Real-Time Models Are Reshaping the Enterprise

Jazper Carter

Cybersoft Publishing LLC

Cybersoft Publishing LLC

First Edition, Jan. 2026

Table of Contents

1 Introduction

1.1 Why This Book Exists

Most digital-twin programs fail at a layer that has nothing to do with technology. The sensors connect, the simulation engine runs, and the dashboard renders. Then the program stalls, not because the platform underdelivered but because no one established who owns the data, who is authorized to act on the outputs, and what standard of quality those outputs need to meet before a consequential decision is made against them. This book is organized to give enterprise leaders both the technical literacy to evaluate and direct digital-twin programs and the management architecture to make those programs durable.

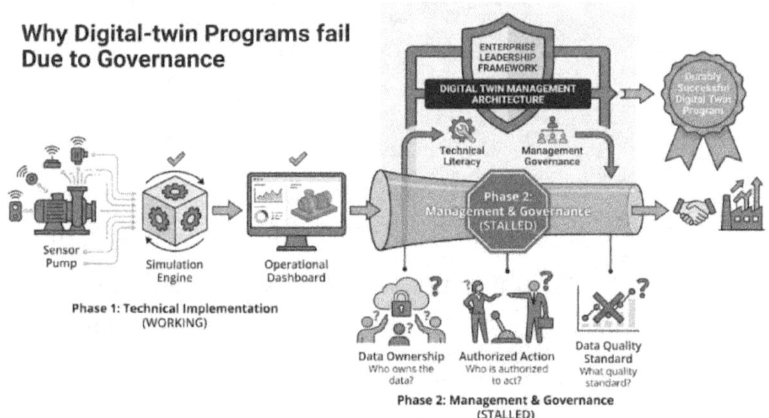

The nine chapters follow a deliberate sequence. Chapters One through Three establish foundational literacy: what a digital twin is, how the technology stack fits together, and where documented operational value has been realized across verticals. Chapters Four through Six address the core management challenges: building a credible business case, designing the data architecture that keeps a twin synchronized, and integrating AI and simulation capabilities. Chapters Seven through Nine address governance, organizational readiness, and the path from pilot to sustained enterprise scale.

Diagram I.1 - The Nine-Chapter Journey Roadmap

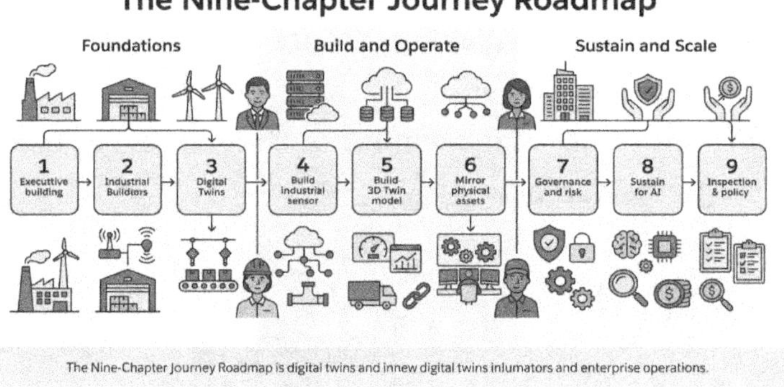

The Nine-Chapter Journey Roadmap is digital twins and innew digital twins inlumators and enterprise operations.

1.2 What Is a Digital Twin

For the purposes of this book, a digital twin is a living digital model of a physical asset, process, or system that is kept in continuous or near-continuous synchronization with its real-world counterpart through data and used operationally for monitoring, simulation, prediction, and optimization. Three elements in that definition are load-bearing. It is a living model: unlike a design schematic or a one-time simulation, a twin updates as conditions change.

It is kept in sync: the data pipeline connecting physical and digital is not optional; it is the mechanism that makes the twin useful for present-state decisions. And it is used operationally: a twin that informs no decision and triggers no action is an expensive display, not a management instrument.

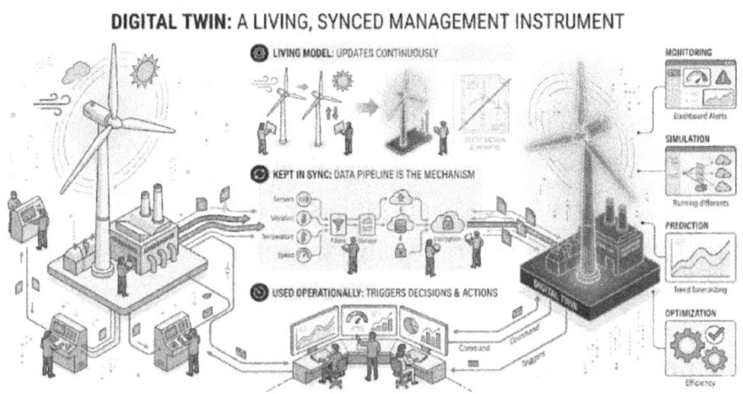

Twins come in several types that differ in scope and complexity. An asset twin models a single piece of equipment. A process twin models a workflow across multiple assets. A system twin models an interconnected set of processes, such as a power distribution grid or a regional supply chain. Most programs begin with asset twins and expand outward as data infrastructure and governance mature. The maturity ladder, from descriptive (what is happening now) through diagnostic, predictive, and prescriptive, to autonomous (closed-loop action), determines what questions a twin can answer and what governance controls it requires. A program that attempts to skip rungs without building the underlying data-quality and model-validation infrastructure will accumulate trust deficits that surface at the worst possible moment.

1.3 The NorthArc Industries Thread

Abstract frameworks become operational when they are applied to a concrete set of decisions. Each chapter of this book follows a recurring cast of leaders at NorthArc Industries, a fictional mid-sized multi-business enterprise headquartered in Columbus, Ohio, with roughly 12,000 employees across three operating units: NorthArc Manufacturing, NorthArc Energy Services, and NorthArc Logistics. NorthArc is fictional; the decisions its leaders face are drawn from recurring patterns in real enterprise programs.

The program was sponsored by Renata Vance, the Chief Digital Officer, following a costly turbine outage in 2024 that exposed the limits of reactive maintenance. Hector Salinas, VP of Operations, is the skeptical line-of-business owner whose Dayton plant pilots became the proof point. Dr. Mei Lin Park, Director of Enterprise Architecture and Data, owns the integration patterns that determine whether twins remain synchronized or drift apart. Dorian Whitlow, Head of AI and Advanced Analytics, governs the predictive and simulation models. Aisha Bramwell, Chief Risk and Compliance Officer, owns audit posture and regulatory engagement. Tomas Reyes, Plant Manager at Dayton, is the operator voice: what appears on the control-room wall and whether runbooks reflect what the twin is reporting. Priscilla Okonkwo leads procurement and vendor management, and her evaluations anchor the vendor-selection discussions throughout the book.

Each chapter opens with a NorthArc scene tied to a specific decision. The arc is not a success story written backward; it includes setbacks and course corrections that reflect how real programs actually evolve. By the final chapter, the reader will have followed NorthArc from initial pilot framing through platform selection, data architecture, AI integration, governance policy development, and program scaling.

1.4 The Manager's and Lead Engineer's Lens

This book is not a developer manual, a vendor comparison report, or an academic survey. It will not teach you to write physics-informed neural networks or configure an OPC UA broker. What it provides is the conceptual vocabulary to have informed conversations with the engineers and vendors who do those things, the management frameworks to direct and govern their work, and the decision structures to protect your organization from failure modes that are well understood in this field but rarely surfaced in vendor presentations.

Every section is written in response to four questions that define the Manager's/Lead Engineer's lens. What does this concept mean for a manager responsible for outcomes but not for implementation? What decision does it inform, and who should own that decision? What risk does understanding it help reduce? What specific action or artifact does it produce, and how does a manager know the action has been taken or the artifact is fit for purpose? These questions keep the discussion grounded in workflow-level impact and operational clarity rather than technical capability for its own sake.

Diagram I.2 - The Manager's Lens: Four Guiding Questions

Each chapter closes with a Manager's Checklist built from these four questions, and a Takeaway that names the single most important decision or artifact the chapter addresses. The checklists are not summaries; they are accountability instruments, pointing to the specific decisions, ownership structures, and artifacts, such as pilot design documents, model risk registers, data ownership matrices, and vendor scorecards, that distinguish programs with operational clarity from programs that exist primarily in slide decks.

Readers new to digital twins will find that the first three chapters build sufficient literacy to engage the remainder of the book without a technical background. Readers with program experience will find that the frameworks in Chapters Four through Nine provide structure for decisions they may have been making instinctively or by precedent. The goal in both cases is the same: to give managers the tools to lead programs that deliver on their promise rather than exhaust their budget proving a point on a pilot that never scales.

2 Not a Dashboard with Better Graphics: The Operational Reality of a Digital Twin

2.1 The Turbine Outage That Started a Conversation

Three weeks after the 2024 turbine outage at NorthArc Energy Services' Westfield generating station, Renata Vance called a working session she described in the meeting notice as a plain-language review of where the company actually stood on digital twin capability. The Westfield event had cost eleven days of lost generation, a six-figure mobilization for repairs, and a regulatory inquiry into whether condition monitoring had been adequate. A lack of data had not caused the outage. NorthArc had sensor feeds, historian data, and a real-time operational dashboard displaying turbine temperatures, vibration signatures, and lube-oil pressures in a visualization suite that the vendor had sold partly on the strength of its appearance.

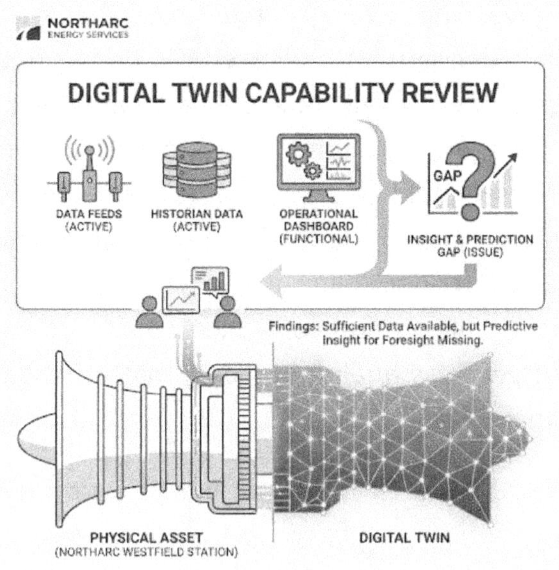

Hector Salinas, VP of Operations for NorthArc Manufacturing, arrived with a pointed question. His engineering team had found that the dashboard had been showing elevated vibration readings in one bearing assembly for at least 9 days before the failure. The readings were accurate. No alert had been generated, no model had projected a failure timeline, and no recommendation had been issued for preventive inspection. The vendor's response: the system was functioning as designed. It monitored and displayed. Simulation, prediction, and decision support were on the product roadmap for a future release.

That working session became the founding conversation for NorthArc's enterprise digital twin program. Vance drew a line governing every purchasing, architecture, and governance decision that followed. On one side were tools that record and display the present state of a physical system. On the other hand, some tools maintain a living, synchronized model of a physical system, use that model to simulate and predict behavior, and feed those predictions into decisions before consequences become irreversible. The difference, Vance concluded, was not a matter of graphics quality. It was a matter of operational architecture.

Diagram 1.1 - The NorthArc Working Session: Dashboard vs. Twin

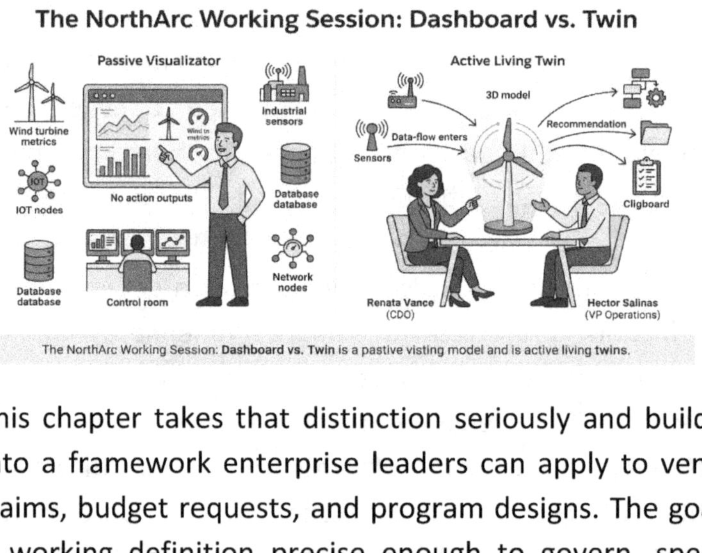

The NorthArc Working Session: **Dashboard vs. Twin** is a pastive visting model and is active living **twins.**

This chapter takes that distinction seriously and builds it into a framework enterprise leaders can apply to vendor claims, budget requests, and program designs. The goal is a working definition precise enough to govern, specific enough to write into vendor requirements, and honest enough to surface the gap between what organizations believe they have purchased and what they have actually deployed. That gap, as NorthArc discovered, has operational consequences.

Throughout this book the NorthArc cast will return in scenarios from different stages of their digital twin journey: Renata Vance making strategy calls, Hector Salinas testing frameworks against manufacturing reality, Dr. Mei Lin Park wrestling with data architecture, Dorian Whitlow managing model lifecycle, Aisha Bramwell anchoring governance, Tomas Reyes translating twin outputs into control-room practice, and Priscilla Okonkwo negotiating with vendors who speak the language of digital twins fluently but do not always deliver them.

2.2 What a Digital Twin Actually Is: Three Non-Negotiable Properties

The term 'digital twin' has been applied to objects ranging from a static CAD file with sensor overlays to a continuously evolving AI-augmented simulation environment capable of generating thousands of predictive scenarios per second. That range is not an academic problem. It is a procurement, governance, and risk problem. When leaders cannot draw a line between what qualifies as a digital twin and what does not, they cannot evaluate vendor claims, set scope expectations, or hold programs accountable. The definition that follows is operational, constructed around three properties that separate a genuine digital twin from the many things that share its name.

Continuous Synchronization: The Living Link Between Physical and Digital

A digital twin is not a model that was accurate when it was built. It is a model that remains accurate because it receives a continuous, automated stream of data from the physical system it represents. The digital model's state at any moment reflects the physical system's actual state within the latency tolerance defined by the use case. For an asset twin of a gas turbine, that means sensor data covering temperature, vibration, pressure, and bearing condition flowing from IIoT (Industrial Internet of Things) edge devices through an edge gateway, over OPC UA or MQTT protocols, into a time-series database such as OSIsoft PI or InfluxDB, and from there into the model's state variables.

A dashboard that reads from the same sensor feeds is not a twin by this criterion because it lacks a state model. It renders a view. Maintaining a state model enables something a dashboard cannot do: the model can be queried, exercised, and run forward in time. Synchronization is what makes a digital twin available for simulation. Without it, the model is a snapshot, and snapshots cannot answer questions about what will happen next. Dr. Mei Lin Park, NorthArc's Director of Enterprise Architecture, put it plainly: 'A dashboard reads from a data store. A twin writes to a model, and the model is a living artifact you can interrogate.' That distinction became the first line in NorthArc's digital twin procurement standard.

Diagram 1.2 - Continuous Synchronization: From Physical Asset to Digital State

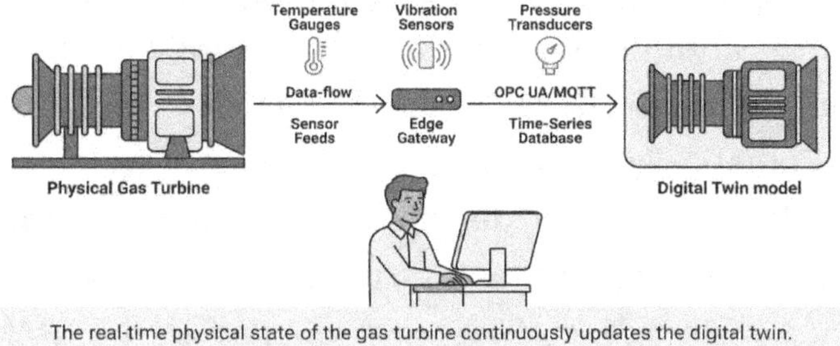

The real-time physical state of the gas turbine continuously updates the digital twin.

Bidirectionality: When the Digital Informs the Physical

A genuine digital twin not only receives information from the physical world; it sends information back. That return path may take the form of alerts, recommendations, setpoint adjustments, or maintenance work orders. The critical point is that the model is not a passive observer. It is an active participant in the operation of the physical system. Bidirectionality operates at different levels depending on twin maturity. At the most basic level, the twin sends a threshold alert. At a more advanced level, it generates ranked recommendations. At the highest level, it issues setpoint changes directly to the physical system's control infrastructure, subject to human approval or, in approved autonomous configurations, without it.

Hector Salinas' plant team encountered this distinction when they piloted a basic twin for a stamping press line at

NorthArc's Dayton facility. The initial configuration was unidirectional: data flowed in, a condition monitor ran, and engineers reviewed output reports twice daily. The reports were accurate and useful, but the value was limited because insights arrived on a reporting schedule rather than in time to prevent the events they described. The shift to bidirectionality meant that model outputs were automatically routed to the maintenance work order system. When the twin detected a lubrication anomaly consistent with a bearing nearing the end of its service life, a planned maintenance ticket was opened in the CMMS before the shift ended. That integration was what Salinas described as the moment the twin became useful.

Simulation Capability: Answering What Would Happen Before It Does

A digital twin can be run forward in time or under hypothetical conditions to generate predictions and explore scenarios that have not yet occurred. This is the property that most sharply separates a twin from a dashboard or real-time monitoring system. A dashboard answers questions about what is happening now. A digital twin answers questions about what will happen and what would happen if conditions changed. Simulation capability requires that the model contain enough physics, operational logic, or learned behavior to project plausible future states. The representation may be physics-based, data-driven, or a hybrid combining physics-based constraints with learned correction factors.

Dorian Whitlow's AI and Analytics team built the simulation layer for the Dayton press line twin using a surrogate model approach. The physics of the stamping process were too complex to simulate from first principles at operational speed, so the team trained a neural network on three years of historical sensor and production data to produce a fast-running proxy that could simulate four hours of future press behavior in under thirty seconds. That simulation speed was the threshold requirement: the operations team needed scenario answers fast enough to make shift-level scheduling decisions. A simulation that ran overnight would not have changed operational practice regardless of its accuracy.

Diagram 1.3 - The Three Non-Negotiable Properties of a Digital Twin

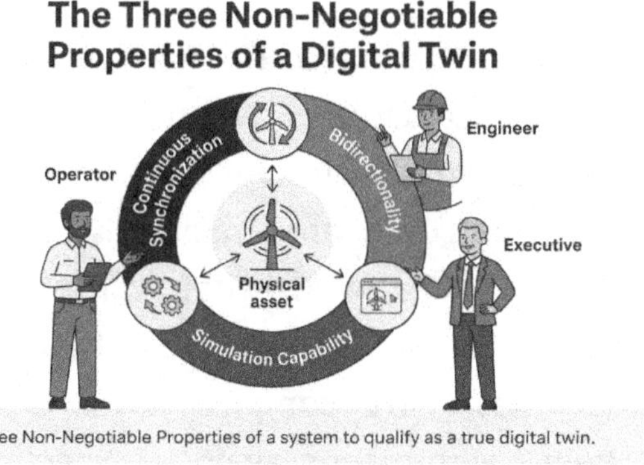

All three Non-Negotiable Properties of a system to qualify as a true digital twin.

2.3 Why a Dashboard Is Not a Twin: The Governance Consequence

The distinction between a dashboard and a digital twin is a governance distinction because the two tools support different decisions and create different accountability structures. A dashboard is a reporting instrument. Its decisions are human decisions made by individuals who observe and exercise their own judgment. The dashboard bears no responsibility for those decisions because it makes no claim about the future and issues no recommendation. When functioning as designed, a digital twin is a decision-support instrument. Its outputs are claims about the physical world that carry predictive content. Those claims can be correct or incorrect, well-calibrated or overconfident. Because a digital twin's outputs are claims, they require governance.

This matters directly to Aisha Bramwell, NorthArc's Chief Risk and Compliance Officer. When turbine monitoring was a dashboard, the accountability structure was clear: engineers reviewed data and made maintenance calls. When NorthArc moves to a digital twin that generates maintenance recommendations algorithmically, the structure becomes more complex. Who is responsible when the twin recommends against inspection and the asset fails? What records must NorthArc maintain to demonstrate that its twin governance met the standard of care required by its operating licenses? These questions have precise answers, but building those answers requires leaders first to understand exactly which instrument category they are deploying.

Diagram 1.4 - Dashboard vs. Digital Twin: Decision Accountability Spectrum

Dashboard vs. Digital Twin:
Decision Accountability Spectrum

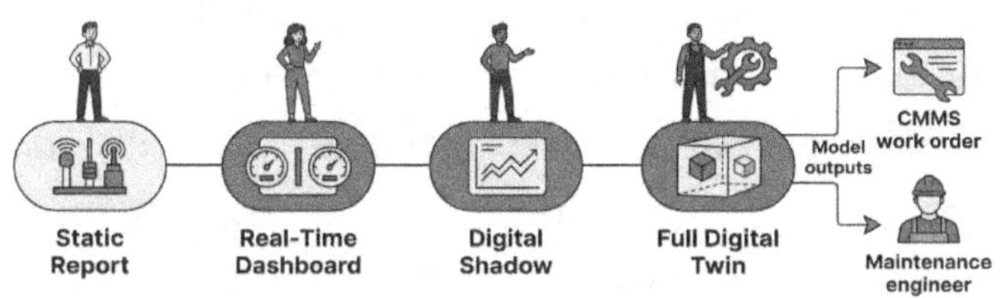

Digital twin, loxes the linest of human-machine and enterprises, operatprise digital twins create accountabilly chains that dashboards do not.

Enterprise leaders who accept vendor claims that a real-time monitoring dashboard constitutes a digital twin are not simply getting a misleading product description. They are accepting a governance mis-framing that will shape every downstream decision: what oversight the program requires, what model risk management applies, what audit trail is needed, and what the organization can legitimately claim to regulators, insurers, and the board. Getting the definition right at the start is a governance act, not a technical one. Renata Vance built a three-property demonstration requirement into NorthArc's vendor evaluation framework: any system proposed as a digital twin had to demonstrate continuous synchronization with a live data source, run a simulated scenario compared to historical ground truth, and route an output to an operational workflow. Vendors who could not complete all three were redirected to the BI procurement track, regardless of how they described their products.

2.4 The Maturity Ladder: Five Levels from Mirror to Autonomous Agent

Digital twins do not arrive fully formed. Organizations build toward twin capability through a sequence of maturity levels, each adding new functional properties to the existing foundation. The five-level maturity model presented here is grounded in the convergence of frameworks from the ISO 23247 manufacturing digital twin standard, the Industrial Internet Consortium reference architecture, and operational vocabulary from production deployments across manufacturing, utilities, and logistics. The levels describe what a twin can do, not what technology it uses.

Level One: Descriptive (The Digital Mirror)

At Level One, the twin is a digital mirror. It maintains a continuously synchronized representation of the physical system's current state. You can query it for the system's current readings, compare them to baseline, and identify anomalies based on threshold rules. What the Level One twin cannot do is explain why an anomaly is occurring, predict whether it will persist, or recommend a response. It is a real-time fact provider. Level One twins are appropriate when the primary operational need is situational awareness at scale, and they form the foundation for higher-maturity levels. NorthArc's

Westfield monitoring, had it been properly architected, would have qualified as Level One. The failure was not that Level One was insufficient; it was that the organization called it something more without building the infrastructure to support the next level.

Diagram 1.5 - Digital Twin Maturity Ladder: Five Levels

Digital Twin Maturity Ladder: Five governance complexity and learn 3D twin models mirrorimg physical assets.

Level Two: Diagnostic (The Root-Cause Engine)

At Level Two, the twin adds diagnostic capability: the ability to analyze current and historical model states to explain why a condition exists. Diagnostic twins use correlation engines, fault trees, and physics-based reasoning to connect an observed anomaly to its probable cause. When the Dayton press shows elevated bearing vibration, the Level Two twin not only flags the anomaly; it cross-references the historical vibration pattern and lubrication records and returns a diagnosis with a confidence score. Diagnostic capability requires a knowledge base of failure modes and cause-and-effect relationships. This is where the first serious data quality debts become visible: a diagnostic engine is only as good as the historical data it reasons against.

Level Three: Predictive (The Failure Forecaster)

At Level Three, the twin adds a forward-looking dimension: the ability to project the physical system's future state from its current state and a model of how the system evolves. Predictive twins answer questions such as: given today's vibration signature and lubrication condition, what is the probability that this bearing will require replacement within the next thirty days? Predictive capability transforms a condition-monitoring tool into a maintenance-planning tool. Dorian Whitlow's team at NorthArc used a hybrid approach for the turbine fleet, combining a physics-based creep-and-fatigue model for hot-section components with a gradient-boosted survival model for balance-of-plant components where failure modes were more heterogeneous and historical data was richer than engineering knowledge.

Level Four: Prescriptive (The Decision Advisor)

At Level Four, the twin generates and ranks recommended actions. A prescriptive twin does not simply forecast that a bearing will fail in thirty days. It recommends a specific maintenance action at a specific time, compares it against alternatives, quantifies the cost and risk for each option, and presents the analysis to the human decision-maker.

Prescriptive capability requires the twin to have models of the decision space and the physical system, encoding which actions are available, their costs, and the outcome probabilities each action produces given the current physical state. A prescriptive twin whose outputs arrive in a report sitting in an engineering inbox for three days is not delivering prescriptive value at the speed Level Four capability is intended to provide.

Diagram 1.6 - Prescriptive Twin Output: Decision Options and Expected Outcomes

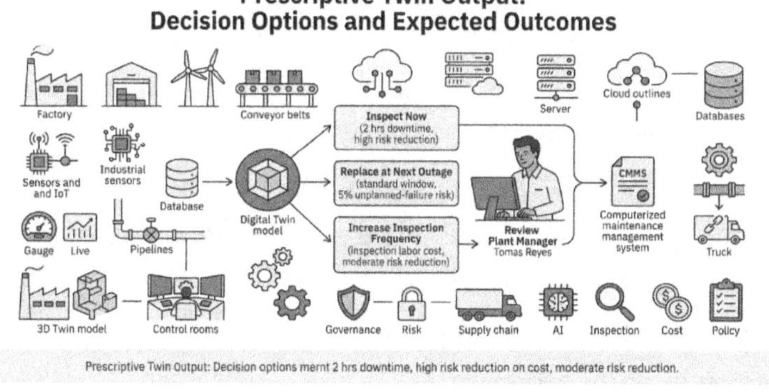

Level Five: Autonomous (The Self-Optimizing System)

At Level Five, the twin closes the loop without human approval for specifically defined classes of decisions. The autonomous twin monitors, diagnoses, predicts, selects an action from its approved decision library, and executes it through a direct interface to the physical system's control infrastructure, within a governance envelope defined by human operators. Autonomous operation is not ungoverned; it is human-approved automation within human-defined boundaries, with a full audit trail and override capability. It is appropriate only for decisions that are well-characterized, low-risk in individual instances, high-frequency, and where the cost of human review exceeds the cost of model error. NorthArc's Aisha Bramwell framed it precisely: 'Autonomous operation is a governance classification, not a technology classification. We decide what the twin is allowed to do without asking. The twin does not get to expand its own authority.'

2.5 Four Twin Types: Asset, Process, System, and Organizational

The maturity ladder describes a twin's level of maturity. The type taxonomy describes what it represents. Enterprise leaders need both dimensions to frame programs correctly, because the right type of twin for a given problem is not always the most obvious one, and mixing type with capability is a common source of scope confusion in early programs.

Asset Twins

An asset twin is a digital twin of a specific physical asset: a turbine, a compressor, a pump, a bridge, or any bounded physical object that can be instrumented and modeled. Asset twins are the most common entry point for digital twin programs in manufacturing and utilities because they have a clear physical boundary, established instrumentation practices, and an existing body of engineering knowledge on their failure modes. Asset twins are valuable for predictive maintenance, remaining useful life estimation, and performance optimization at the asset level. Their limitation is that they optimize the asset in isolation, which may not align with optimizing the process or network to which the asset belongs.

Process Twins

A process twin models an end-to-end operational process, such as a production line, a logistics flow, a supply chain sequence, or a service delivery process. Where an asset twin asks what this asset is doing and will do next, a process twin asks how this end-to-end process is performing and where its constraints lie. NorthArc Logistics deployed a process twin for its inbound freight coordination process after Westfield revealed that parts availability had extended the repair by four days. The process twin ingested carrier-tracking, warehouse-management, and supplier-production data to maintain a real-time model of the parts pipeline. When a shipment was delayed, the team ran scenario analysis of alternative sourcing options and flagged the recommendation before the original delivery date was missed.

System Twins

A system twin models the behavior of an interconnected network of assets and the relationships among them. The classic example is an electrical grid twin, which models not just individual transformers and transmission lines but the network topology that connects them and the flow equations governing how power moves through the network under different demand and supply conditions. A system twin can answer questions that neither an asset twin nor a process twin can answer: if this transmission line goes out of service, what is the contingency loading on adjacent lines, and which is closest to its thermal limit? Dr. Mei Lin Park's architecture team built a reference architecture for NorthArc's grid twin that started with Level One synchronization of all monitored assets, added Level Two diagnostic capability, and reserved the system-level network flow simulation for a third phase requiring model validation against historical outage data.

Diagram 1.7 - Four Twin Types: Asset, Process, System, Organizational

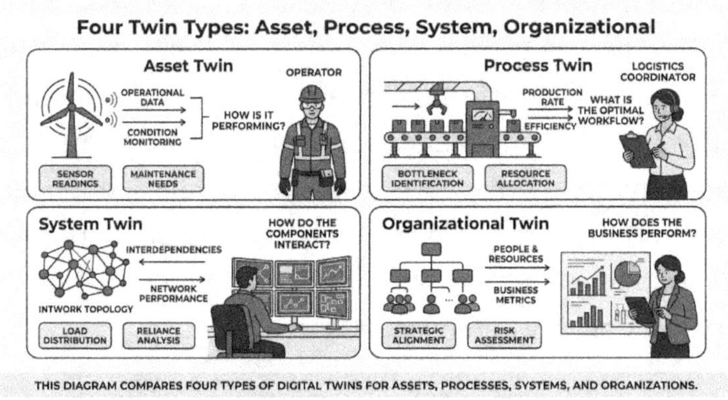

THIS DIAGRAM COMPARES FOUR TYPES OF DIGITAL TWINS FOR ASSETS, PROCESSES, SYSTEMS, AND ORGANIZATIONS.

Organizational Twins

An organizational twin models not a physical asset or process but an organization's operational capacity: workforce availability, skill distribution, resource allocation, and the match between organizational capability and operational demand. Organizational twins are built from HR systems, time-and-attendance data, training records, and operational demand forecasts. They are used for workforce planning, capacity management, and scenario modeling of organizational change. Most enterprise programs exclude organizational twins from initial scope for good reason: the required data integration across HR, operations, and finance is technically complex and politically sensitive, and workforce privacy questions demand careful governance design before any deployment.

2.6 Common Misconceptions That Managers Must Correct

Digital twin programs accumulate misconceptions at each stage of the enterprise decision chain: vendor sales conversations, internal program proposals, executive briefings, and governance reviews. Managers responsible for these programs must identify and correct the most consequential misconceptions before they become embedded in program design.

The first misconception is that a digital twin is primarily a visualization. The twins' visualization layer is the most visible part of the system and often the feature most prominently demonstrated in vendor sales cycles. But the visualization is the output layer. The twin is the model that produces it. Organizations that invest heavily in the visualization layer while underinvesting in synchronization infrastructure, data quality management, and simulation capability will have an impressive screen that is as operationally limited as the Westfield dashboard.

The second misconception is that higher fidelity is always better. The required fidelity of a digital twin is determined entirely by the decisions it is intended to support. A twin used to schedule preventive maintenance does not need to model the aerodynamic boundary-layer behavior of turbine blades. Building and maintaining unnecessary fidelity wastes modeling resources, increases integration complexity, and degrades real-time performance. Priscilla Okonkwo, NorthArc's procurement lead, added a specific question to every twin vendor evaluation: 'What is the minimum fidelity required to answer the business question, and how did you derive that requirement?' Vendors who could not answer precisely were revealing that they were optimizing for demonstration quality rather than operational fit.

The third misconception is that a digital twin program is a technology project. It is not. It is an operational change program that uses technology as an enabling layer. The most technically excellent digital twin produces zero operational value if maintenance planners do not trust its outputs, if its recommendations are not routed into the workflows where decisions are made, or if model predictions are never reviewed for accuracy. Tomas Reyes, plant manager at Dayton, put it plainly in a steering committee update: 'My team will use what we trust. We will trust what we can verify. Build me a twin I can verify, and I will build you a team that uses it.' That comment led NorthArc to add a model-transparency requirement and a control-room verification protocol that were not in the original plan.

Diagram 1.8 - Three Misconceptions and Their Operational Consequences

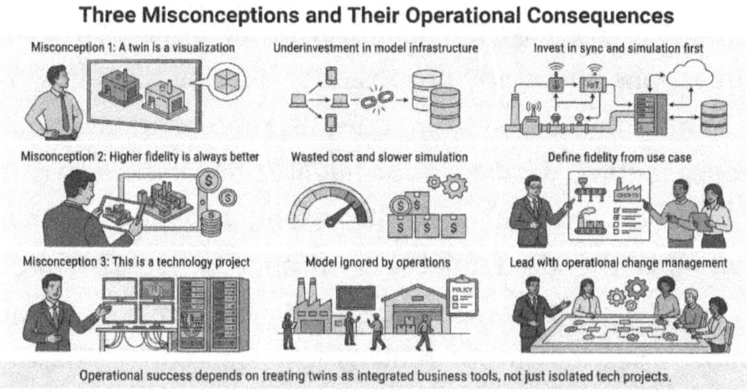

The fourth misconception is that digital twins replace human judgment. They do not. At every maturity level, the twin operates within a governance envelope defined by human decisions about what it is allowed to do. The twin amplifies human judgment by providing better information, faster analysis, and quantified uncertainty. Leaders who position digital twins as judgment-replacing rather than judgment-supporting systems will encounter resistance from the frontline operators and engineers whose expertise and buy-in are essential to making recommendations operationally effective.

2.7 What Enterprise Leaders Must Demand: Four Governance Questions

The operational definition of a digital twin, grounded in the three properties of continuous synchronization, bidirectionality, and simulation capability, provides the foundation for a governance definition that enterprise leaders can use in vendor requirements, program charters, and portfolio oversight. The governance definition asks four questions about any proposed or existing digital twin. First: What physical system does it represent, at what boundary, and at what level of fidelity? Second: How is it synchronized with the physical system, at what latency, and what is the data quality assurance process? Third: What decisions is it intended to support, and through what workflow does its output reach those decisions? Fourth: How is its performance tracked, and what is the process for identifying and correcting model drift?

Those four questions distinguish programs that are genuinely governed from programs that have governance documents but lack governance practice. A program that cannot answer the first question does not have a defined scope. A program that cannot answer the second does not know whether its model is accurate. A program that cannot answer the third is not producing operational value. A program that cannot answer the fourth is operating on a model confidence that may be unjustified. Renata Vance built these questions into NorthArc's quarterly digital twin steering committee review template.

Each active twin program was required to provide current answers at every steering meeting. The discipline proved more valuable as a program management mechanism than any project plan the team had tried, because it forced teams to maintain operational awareness of their twin's actual performance rather than tracking against a project milestone list.

Leaders must also demand that vendors distinguish clearly between what the product does today and what the roadmap describes for future releases. The market for digital twin platforms is projected to reach approximately $25 billion by 2030, and with that market size has come the predictable pattern of vendors who describe their products in the vocabulary of their aspirational future state. Leaders who require vendors to demonstrate capability against specific test cases using the buyer's own data before contract signature will avoid a significant proportion of the disappointments that characterize first-generation digital twin programs.

Diagram 1.9 - The Four Governance Questions for Digital Twin Programs

The Four Governance Questions for Digital Twin Programs

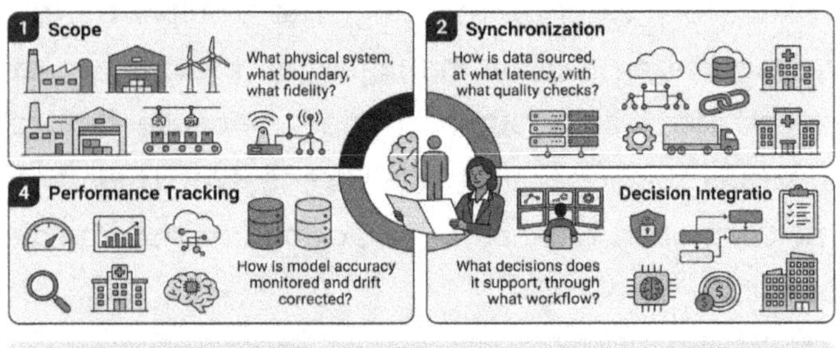

The digital twin shamp ıtelies na: the Four governance questions for digital twin programs.

2.8 The Operational Reality: Twins in Regulated and High-Stakes Environments

The NorthArc scenarios in this chapter are drawn from manufacturing, energy services, and logistics. Still, the operational definition of a digital twin applies equally in regulated sectors where the stakes of definitional confusion are higher. Healthcare facility operations, utility grid management, oil and gas asset integrity, and defense logistics all have active digital twin deployments. All have encountered the same pattern: organizations that purchased monitoring and visualization systems under the twin label and discovered, at the moment of an adverse event, that the system lacked the simulation and decision-support capability the twin label had led them to expect.

In regulated environments, the definition of governance takes on additional dimensions. A healthcare system deploying a hospital operations twin must determine whether the twin's recommendations constitute clinical decision support subject to FDA oversight or operational decision support subject to standard software quality management. An electric utility operating a grid twin under FERC reliability standards must document how the twin's recommendations relate to the operator's independent obligation to assess system conditions. An oil and gas operator using an asset integrity twin must demonstrate to its regulator that the twin's inspection interval recommendations are supported by a validated degradation model rather than pattern matching against insufficient historical data.

These compliance dimensions do not change the fundamental definition of what a twin is. They do change what it takes to operate a twin responsibly in these environments. Leaders in regulated sectors must add a fifth governance question to the four listed in the preceding section: How does this twin's output relate to the regulatory obligations that govern the decisions it is designed to support? Answering that question requires collaboration between the digital program office and the organization's regulatory affairs and legal teams from the start of the program, not after a regulator raises the question in an audit. NorthArc's Energy Services unit engaged with its regional grid operator and regulatory counsel when scoping the grid twin specifically to establish whether the twin's contingency analysis outputs would be treated as operator decision aids or as formal reliability studies. That answer, which took three months to obtain, shaped the entire architecture of the twins' output layer.

2.9 Manager's Checklist: Early Framing for a Digital Twin Program

The decisions below represent the minimum framing actions a manager should complete before committing budget or naming a system as a digital twin. Completing them prevents the most common and costly errors in first-generation programs.

- Apply the three-property test to any system described as a digital twin: Does it maintain continuous synchronization with the physical system? Does it have a bidirectional return path from digital to physical? Does it have a simulation capability that can project future states or answer what-if questions? A system that fails any one of these tests is not a digital twin.
- Distinguish clearly between the twin type you need and the maturity level you can realistically deliver in the first program phase. Naming an asset twin at Level One maturity as the first deliverable is honest scoping. Naming a Level Four system twin as the program outcome without a funded pathway to get there is scope inflation.
- Build the four governance questions into your program charter before kickoff: scope definition, synchronization architecture, decision integration pathway, and performance tracking process. These questions should be answerable at launch and re-answerable at every quarterly review.

- Require vendors to demonstrate the three non-negotiable properties against your data and your use case before contract signature. A vendor demo on curated demonstration data does not prove the system will perform on your operational data.
- Engage your legal and compliance team in program framing from the start, not after the architecture is set. In regulated sectors, the regulatory classification of the twins' output layer may constrain design choices that are expensive to reverse after implementation.
- Map the operational workflows where the twins' outputs must land before designing the model architecture. A recommendation that cannot be acted on within the timeframe of the decision is intended to support, but delivers no operational value, regardless of its technical accuracy.
- Define the performance tracking process as part of program design. Identify the metrics for model accuracy, the frequency of performance reviews, the threshold that will trigger recalibration, and the team responsible for carrying out that process. A twin without a performance-tracking plan will drift without detection.
- Do not allow the visualization layer to consume a disproportionate share of program investment. Visualization is important for adoption and usability, but it is the output layer. Investment in data synchronization infrastructure, data quality

management, and model validation delivers more operational value per dollar.

- Establish a model governance log as an operational artifact from the first deployment. Record design decisions, training data, validation results, performance metrics, and any post-deployment changes. This log is the audit trail that supports accountability when model outputs are used in consequential decisions.
- Communicate the program's maturity level honestly to senior stakeholders. A Level One twin is valuable. It is also not a predictive maintenance solution. Managing executive expectations about what the current investment delivers, and what reaching the next maturity level will require, is a governance obligation for program managers.

2.10Takeaway

The distinction between a digital twin and a dashboard is not semantic. It is the difference between a system that tells you what is happening and a system that tells you what will happen and what to do about it. Enterprise leaders who accept loose definitions at program start will find those definitions propagate through every subsequent decision: vendor selection, architecture choices, governance design, and operational integration. The three non-negotiable properties, continuous synchronization, bidirectionality, and simulation capability, provide a precise filter that separates genuine digital twins from the more numerous systems sold under that label. The maturity ladder maps the investment and capability-building path from a digital mirror to a decision-support system to an autonomous optimization agent. The four twin types provide the scope of vocabulary needed to define what the program will model and why.

Renata Vance's working session after Westfield produced one conclusion worth carrying into every chapter that follows: the question is never whether a digital twin is technically impressive. The question is whether it changes a decision that needs to change, before the consequence that motivated the program becomes irreversible. That criterion, workflow-level impact on consequential decisions, is the standard against which every digital twin investment should ultimately be measured. The chapters ahead will show how NorthArc and the broader enterprise community have been built, governed, and measured against that standard across manufacturing, energy, logistics, and beyond.

3 The Architecture of Awareness: Sensors, Data Pipelines, and the Real-Time Backbone

3.1 Walking the Floor Before Touching the Platform

Dr. Mei Lin Park arrived at the Dayton facility on a Tuesday morning, carrying a tablet loaded with NorthArc's network topology diagram and a healthy skepticism that any of it was accurate. Tomás Reyes met her at the main entrance, and they walked the floor together for the next three hours, conducting a sensor coverage audit across six high-speed conveyor drives, two hydraulic press systems, four HVAC chillers, and the main compressor bank.

Tomás was candid about the fact that the instrumentation had been installed piecemeal over fifteen years by different contractors. The network topology on Mei Lin's tablet was missing two edge gateways and showed a vibration sensor array as active, even though it had been decommissioned. She returned to Columbus with a coverage gap map, a list of twelve sensors that were absent, miscalibrated, or intermittent, and questions about which protocols were actually in use versus which ones appeared in vendor documentation. The walk reinforced a principle she would repeat throughout the program: you cannot design an accurate digital twin from documentation alone. The physical world has to be seen, measured, and mapped before a single line of pipeline code is written. This chapter follows the architecture she and the NorthArc team built, layer by layer, from the plant floor to the twin platform.

3.2 The Sensor Stack: IIoT, Measurement Types, and What Managers Must Ask

A digital twin is only as accurate as the physical measurements flowing into it. Every insight the twin produces originates as a sensor reading. Managers who treat the sensor layer as a procurement detail rather than an architectural decision will pay for that choice in the form of data quality debt for years.

The industrial sensor landscape divides into five core measurement families. Vibration sensors capture the mechanical signature of rotating equipment; deviations from baseline frequency profiles indicate bearing wear long before a failure becomes audible. Temperature sensors monitor heat profiles across motors and bearings. Pressure sensors track fluid and gas system states and are foundational to any process twin covering hydraulics or pneumatics. Flow meters measure volume or mass throughput and are critical for energy and yield calculations. Computer vision cameras detect surface defects and verify assembly sequences. At the same time, LiDAR extends spatial awareness into three dimensions, enabling a spatial twin to mirror asset positions and inventory flows in real time.

Diagram 2.1 - The IIoT Sensor Stack

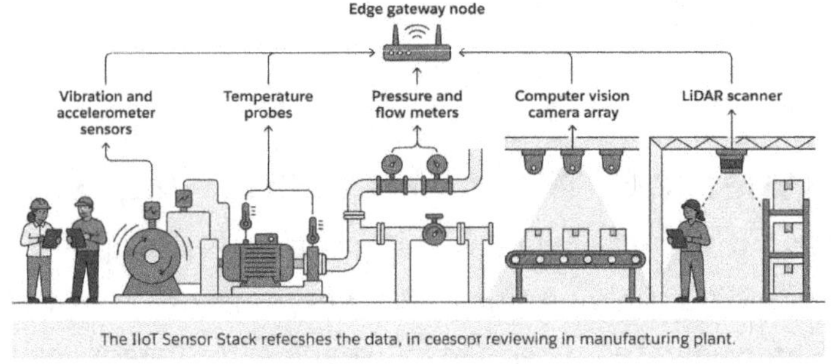

Calibration is the discipline managers most underestimated. A sensor that has drifted from its calibrated baseline does not fail obviously; it continues to produce data that no longer reflects physical reality. The twin ingests that data and begins generating predictions based on corrupted inputs. Mei Lin's team found four temperature sensors at the Dayton facility that had not been recalibrated in over two years, despite a manufacturer's specification for annual recalibration. None were flagged in the asset management system as out of compliance. Managers should require a calibration register as a deliverable at the start of any sensor audit phase, documenting the last calibration date, interval, method, and responsible party for each sensor in scope. Any sensor overdue for calibration is a known data quality risk that will compromise the twin before it is ever turned on.

Sampling Rate: When 1 Hz Is Enough and When You Need 1 kHz

The sampling rate question is a business decision dressed in engineering clothing. For most temperature and pressure monitoring use cases, measurements every one to five seconds are more than adequate because the physical processes are slow. Bearing temperature rising toward a failure threshold moves over minutes or hours. For vibration monitoring on high-speed rotating equipment, the picture is entirely different. A motor shaft spinning at 3,600 rpm has a fundamental rotation frequency of 60 Hz, and its bearing defect frequencies may appear between 100 and 500 Hz. Capturing those signatures requires sampling rates from 1 kHz to 10 kHz.

Computer vision and LiDAR add their own velocity considerations. A camera at 30 frames per second produces roughly 30 megabytes of raw image data per second before compression. LiDAR at 10 scans per second generates similar volumes. Both require purpose-built edge processing that reduces data volume close to the source before anything reaches the enterprise network. Every sensor category has its own data velocity profile, and a single ingestion architecture cannot serve all of them without explicit design choices about where processing happens and how aggressively data is compressed at the edge.

3.3 Edge Gateways and Edge Computing: Processing Where the Data Lives

An edge gateway is a computing device deployed close to the physical assets it serves, within the plant or at the field equipment location, that performs initial data processing before forwarding results to a central system. Understanding what edge gateways do, and what they do not do, is essential for any manager approving a digital twin architecture.

At the most basic level, an edge gateway performs protocol translation. Factory-floor equipment communicates over a variety of industrial protocols, many of them decades old, that are not natively compatible with enterprise IT networks or cloud ingestion pipelines. OPC UA (Unified Architecture) is the modern standard for industrial data exchange. However, many legacy devices still speak Modbus, DNP3, or proprietary protocols, requiring a gateway to translate their signals into a format that upstream systems can process. Mei Lin's audit revealed that the Dayton facility had equipment on three different legacy protocols in addition to OPC UA. This mix would have been invisible in a purely documentation-based architecture review.

Diagram 2.2 - Edge Gateway Architecture: From Sensor to Network

Edge Gateway Architecture: From Sensor to Network

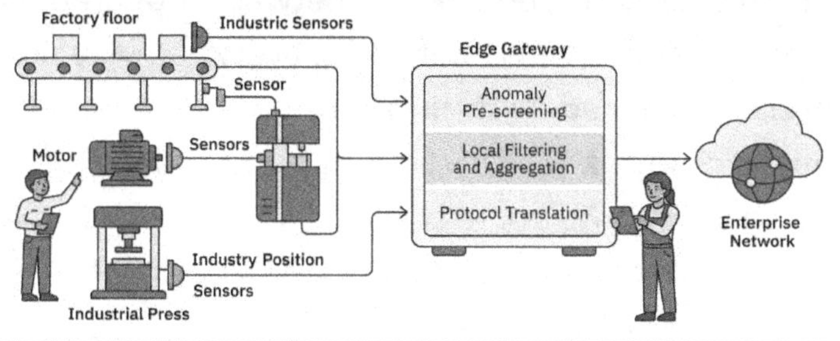

Edge Gateway Architecture: From sensor to network, Anagalnal filtering from PhrEnterprise Network.

Beyond protocol translation, capable edge gateways perform local filtering and aggregation. Rather than forwarding every raw sample to the cloud, the gateway computes statistical features from each sampling window. It forwards only a small number of derived values, rather than thousands of raw points. A vibration sensor sampling at 5 kHz may produce 5,000 data points per second per channel; the gateway reduces that to four computed features per second, a factor of one thousand reduction while preserving the signal information needed for anomaly detection. This edge-based feature extraction is a necessity, not an option, for deployments with many high-frequency sensors.

Edge gateways also provide operational resilience and a critical security boundary. A gateway with local buffering and store-and-forward capability continues collecting data when the WAN link is down, queues it locally, and synchronizes with the upstream platform once connectivity is restored. That capability is also valuable for audit and compliance scenarios requiring continuous sensor data capture. Equally important, the edge gateway sits at the OT/IT boundary between operational technology networks and IT networks. Every gateway must be explicitly hardened, with firmware up to date, default credentials changed, and access controls documented, before production connectivity is approved. An unhardened gateway is not a security enhancement; it is a potential entry point into the OT network.

3.4 Communication Protocols: OPC UA, MQTT, and Kafka

Three protocols dominate the conversation about digital twin data pipelines, and each operates at a different layer of the stack. Understanding where each belongs helps managers evaluate architecture proposals and avoid treating all three as interchangeable options rather than complementary components in a layered design.

OPC UA is a field-level and plant-level communication standard developed by the OPC Foundation and widely adopted across industrial automation vendors. It provides a structured, semantic data model in which each data point is described with context: what it measures, the unit it uses, the normal range, and the asset to which it belongs. That semantic richness means the twin platform receives structured information that maps directly into the twin's asset model, not just a stream of unlabeled numbers. OPC UA also provides built-in security through certificate-based authentication and encrypted transport, making it the preferred protocol for new instrumentation and modern PLC and SCADA integrations.

Diagram 2.3 - Protocol Layers: OPC UA, MQTT, and Kafka in a Digital Twin Pipeline

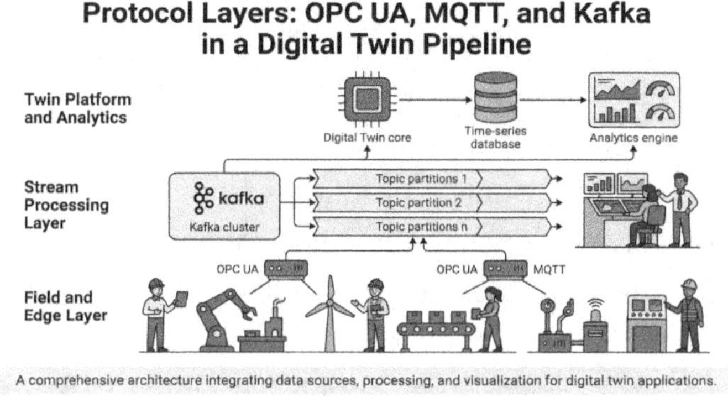

A comprehensive architecture integrating data sources, processing, and visualization for digital twin applications.

MQTT (Message Queuing Telemetry Transport) is a lightweight publish-subscribe protocol for constrained devices and unreliable networks. Where OPC UA is rich and structured, MQTT is lean and flexible. A vibration sensor that cannot support the full OPC UA stack can typically support MQTT. The tradeoff is that MQTT messages carry no inherent semantic context; the meaning of the payload must be agreed upon at design time and enforced through schema management. That discipline requirement is where many deployments create technical debt. MQTT connections proliferate without schema governance, and the pipeline team eventually faces a collection of undocumented payload formats requiring manual reverse-engineering. Kafka operates at a different level entirely. Apache Kafka is a distributed event streaming platform designed to handle very high data volumes with guaranteed delivery, persistent storage, and the ability to replay historical event streams. Where OPC UA and MQTT get data off the device, Kafka moves that data at scale within the enterprise architecture, routing it to multiple downstream consumers simultaneously. A digital twin platform, a time-series database, an anomaly-detection service, and an executive dashboard may all require the same sensor stream. Kafka serves as the single source of truth, decoupling producers from consumers so each can evolve independently. The practical architecture for most enterprise twin deployments runs OPC UA or MQTT from devices to edge gateways, normalizes and buffers at the edge, then publishes into Kafka for distribution. Managers reviewing vendor proposals should be concerned if the proposed architecture omits this stream processing layer.

3.5 Time-Series Databases and Historians: Storing State at Operating Speed

Sensor data is fundamentally temporal. Every reading is a value at a specific point in time, and the analytical value almost always depends on the sequence and timing of readings as much as the readings themselves. A vibration amplitude that increases gradually over three weeks tells a different story than one that jumps overnight. The storage architecture for a digital twin must be designed specifically for temporal queries: compute the average value of this sensor at five-minute intervals over the past ninety days, or find all moments in the last year when this bearing temperature exceeded a threshold within thirty seconds of a vibration spike. Time-series databases such as InfluxDB and TimescaleDB return those queries in milliseconds at production data volumes. A general-purpose relational database returns them in seconds or minutes, if at all.

Diagram 2.4 - Time-Series Data Flow: From Historian to Digital Twin State Model

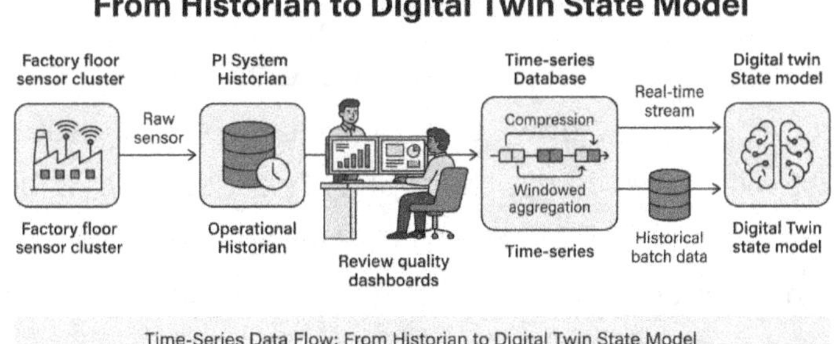

Time-Series Data Flow: From Historian to Digital Twin State Model

The OSIsoft PI System, now part of AVEVA, has been the dominant industrial data historian for over three decades and is present in the majority of large manufacturing, energy, and utilities facilities. For any enterprise building a digital twin where a PI System exists, the historian integration question is not optional. PI data is where the historical baseline lives, and historical baselines are what make it possible to train predictive models and calibrate the twins' state representations. Mei Lin's architecture included a dedicated PI System integration adapter that pulled historical data into the twins' time-series layer during the initial backfill, establishing baseline behavior profiles for each asset before the real-time pipeline was activated.

Managers should ask three questions about time-series storage in any twin proposal. First, what is the designed ingestion rate in data points per second, and how does it compare to the actual sensor count and sampling rates? Second, what is the data retention and downsampling policy, covering how data is aged from high-resolution storage to lower-resolution archive? Storing every raw 1 kHz vibration sample forever is economically impractical, and a defined downsampling policy should be part of the architecture. Third, how does the time-series system integrate with the analytical layer? The query interface between the time-series store and the analytics platform determines how easily new predictive models can be developed and updated.

3.6 The Data Pipeline from Sensor to Twin: End-to-End Architecture

With the individual components described, it is useful to trace the complete path that a physical measurement follows from sensor to twin state update. The flow begins at the sensor, which converts a physical measurement to a digital output. The edge gateway receives that signal, translates the protocol if necessary, applies local filtering or feature extraction, and publishes a structured message to the enterprise message bus. The message bus, typically a Kafka cluster, routes the message to multiple downstream consumers based on topic subscriptions: the time-series database for persistent storage, the twin platform's real-time state-update engine, and an anomaly-detection service that evaluates each incoming measurement against a model of expected behavior.

Three failure modes are most consequential along this pipeline. Sensor dropout, where a device stops transmitting or produces invalid readings, creates gaps in the twin's state record, which can lead predictive models to make incorrect inferences. A model that sees no recent data for a bearing may infer that the asset is healthy when the sensor has failed, and the bearing is operating unmonitored. Pipeline latency spikes introduce temporal inconsistencies into the twin's state model, producing a composite of different moments rather than a coherent snapshot. Schema drift, where a sensor or gateway firmware update changes the message format without updating the pipeline configuration, causes downstream consumers to fail silently or parse data incorrectly.

Diagram 2.5 - End-to-End Data Pipeline: Physical Asset to Twin State

End-to-End Data Pipeline: Physical Asset to Twin State

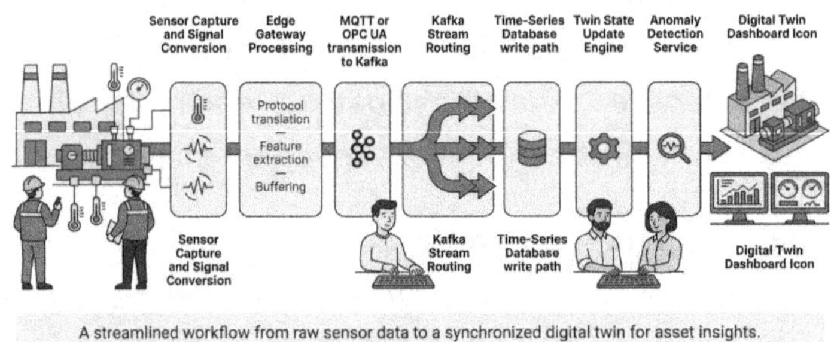

A streamlined workflow from raw sensor data to a synchronized digital twin for asset insights.

Managing these failure modes requires monitoring instrumentation at every stage of the pipeline, not just at the output. Mei Lin's architecture included a dedicated data pipeline health dashboard that tracked message throughput, delivery latency, schema validation failures, and sensor heartbeat status for every data source. That dashboard was not a tool for engineers alone. Tomás and his operations team had read access and a defined escalation procedure for any data quality alert that persisted for more than 15 minutes. Making data quality visible to the people who own the physical assets it represents is a governance choice that produces faster resolution than routing all pipeline alerts through an IT help desk.

Latency Tiers: Matching Pipeline Speed to Business Need

A useful framework divides use cases into three latency tiers. Operational safety and near-real-time control use cases require latency under one second and demand dedicated low-latency pipelines with edge-local processing that does not depend on WAN connectivity. Predictive maintenance and process optimization use cases typically require a latency of 1 to 60 seconds; the early warning signature of a bearing failure evolves over minutes, and a pipeline delivering updates every 30 seconds is entirely adequate. Strategic planning, scheduling, and batch analytics use cases require latency of minutes to hours and are well-served by periodic batch ingestion patterns.

Building a streaming pipeline for a use case that belongs in the third tier delivers no business benefit and adds cost and complexity. Conversely, relying on a batch pipeline for a safety-critical use case is a risk that no cost savings justify. The latency tier classification should be a formal deliverable agreed upon by business stakeholders and the technical team before the pipeline architecture is designed. Latency requirements that live only in the architect's head cannot be audited, changed, or handed off when the program transitions teams.

3.7 Data Quality Debt: The Hidden Cost That Compounds Over Time

Data quality debt is the cumulative effect of unresolved data quality problems in a production pipeline. Like technical debt in software development, each uncorrected problem accrues interest in the form of increasingly unreliable twin outputs, increasingly expensive remediation, and increasingly eroded user trust. In the digital twin context, data quality debt is insidious because it typically accumulates slowly, below the threshold of any single visible failure, and becomes apparent only when a consequential prediction turns out to be wrong.

The most common sources of data quality debt in industrial twin deployments are sensor calibration drift (sensors producing systematically biased readings), undocumented missing data fills (placeholder values inserted during outages that look like real measurements to downstream models), unit inconsistencies (a temperature sensor reading in Fahrenheit ingested as if it were Celsius), and tag naming collisions in the historian (two assets sharing a sensor tag identifier because the historian was not updated when equipment was replaced).

Diagram 2.6 - Data Quality Debt: Sources, Accumulation, and Twin Impact

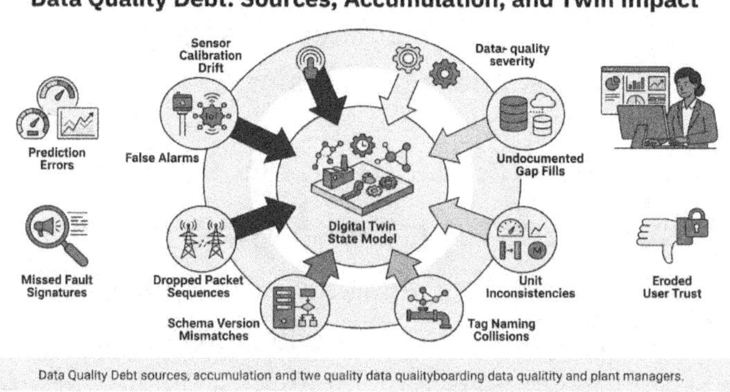

Data Quality Debt sources, accumulation and twe quality data qualityboarding data qualitity and plant managers.

Mei Lin's team developed a data quality scorecard for the Dayton pilot tracking five metrics for each sensor source weekly: completeness (percentage of expected readings received), timeliness (readings arriving within SLA), accuracy (readings passing range and rate-of-change validation), consistency (agreement between redundant sensors), and freshness (time since the most recent valid reading). Any sensor source scoring below the threshold in completeness or accuracy was blocked from contributing to predictive model training until the issue was resolved. The scorecard was also a management tool. It created accountability for sensor ownership, which in an industrial setting is often ambiguous: operations teams own the physical devices, IT teams own the network, and the digital twin program owns the pipeline. By making the scorecard a standing agenda item in cross-functional meetings, Mei Lin ensured that a named individual was assigned to every data quality issue within twenty-four hours. That single practice, requiring no technology investment, was one of the highest-impact governance decisions of the entire pilot.

3.8 Integration Patterns Across Heterogeneous Sources

No enterprise, at any scale, has a homogeneous sensor and data environment. The NorthArc Dayton plant reflects what managers encounter in practice: fifteen years of incremental instrumentation by different contractors, multiple automation vendors, and equipment from different eras with different communication capabilities. Building a digital twin in this environment requires not a single integration pattern but a portfolio of patterns matched to the characteristics of each data source.

The four integration patterns that cover the majority of industrial twin deployments are real-time streaming, periodic batch processing, event-driven triggers, and manual data ingestion. Real-time streaming via OPC UA or MQTT through a Kafka pipeline handles continuous high-value readings, such as vibration, temperature, pressure, and flow. Periodic batch integration handles systems that produce data at defined intervals, such as ERP production records, quality management system results, and maintenance work order completions. Event-driven trigger integration handles alarm activations, equipment state changes, and manual overrides. Manual inspection ingestion covers structured data entry by field technicians that has no automated pathway but contains information the twin's state model needs.

Diagram 2.7 - Integration Pattern Portfolio: Four Pathways to the Twin

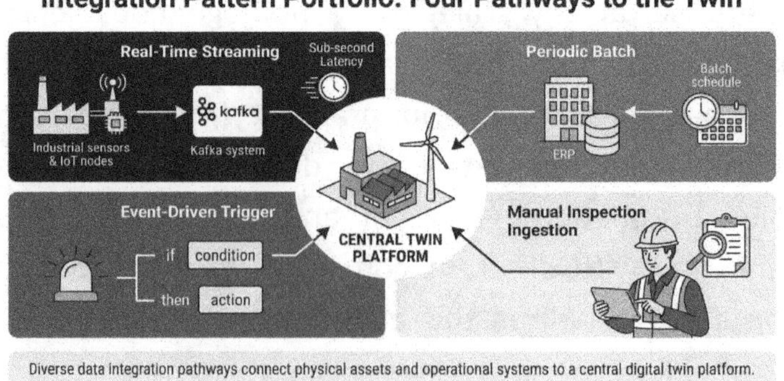

The integration tax is the aggregate cost of connecting and maintaining all of these data pathways. Enterprises frequently underestimate it, focusing budget on the twin platform while treating integration as an implementation detail. Mei Lin's team explicitly calculated the integration tax as part of the Dayton pilot budget, classifying each data source by integration type and assigning a recurring maintenance cost based on the source system's volatility. Integration work represented 38 percent of the total program cost in year one and 22 percent of the ongoing total cost of ownership. Having those numbers documented in advance changed the conversation with NorthArc's finance team from a one-time capital expenditure discussion to a recurring operational cost discussion, which more accurately reflected the program's economics.

Legacy system integration deserves special attention. Every enterprise has at least one system whose integration requires archaeology: reading original vendor documentation, tracing undocumented customizations, and reverse-engineering data formats that were never formally specified. The manager's role is to prevent archaeology projects from becoming indefinite in scope. When an integration team reports that a legacy system's data format is undocumented, the decision framework should be clear: is this data source high-enough value to justify a defined, time-boxed archaeology effort with a specific deliverable, or should it be deferred until the legacy system is replaced? Archaeology without a time box becomes a cost sink.

3.9 Simulation Engines and the AI Layer: From Data to Intelligence

The sensor and pipeline architecture described in the preceding sections produces a continuously updated state representation of the physical world. That state representation is necessary but not sufficient for the analytical value that justifies investment in digital twins. Monitoring the current state is what dashboards do. A digital twin earns its distinction by reasoning forward from the current state to predict future conditions, simulate counterfactual scenarios, and recommend actions. That reasoning capability requires a simulation engine and, increasingly, an AI layer that augments or accelerates physics-based simulation.

Physics-based simulation models encode the mathematical relationships that govern the behavior of a physical system. A thermal simulation model for an electric motor encodes the heat-transfer equations that relate current, load, ambient temperature, cooling airflow, and insulation characteristics to winding temperature over time. Given current-state inputs from the sensor pipeline, that model can project the motor's thermal trajectory and estimate when it will exceed a threshold that triggers a protective shutdown. Physics-based models are accurate within their design envelope, interpretable to engineers who understand the underlying physics, and trustworthy in scenarios that are rare enough to have little or no historical precedent.

Diagram 2.8 - Twin Intelligence Layer: Physics Models, ML Models, and AI Augmentation

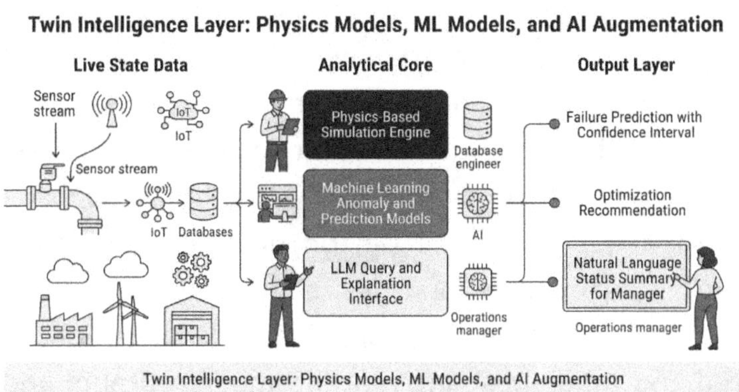

Twin Intelligence Layer: Physics Models, ML Models, and AI Augmentation

Data-driven models learn behavioral patterns from historical data. A machine learning model trained on two years of vibration and temperature histories for the Dayton conveyor drives can learn the multivariate signature that historically preceded bearing failures by a predictable interval. Their weakness is behavior outside the training distribution: they are unreliable predictors for failure modes that have not occurred in the historical record. A hybrid architecture that uses physics models for novel conditions and data-driven models for high-frequency, well-sampled conditions delivers the best of both approaches in most enterprise deployments.

The integration of large language model interfaces into twin context represents an emerging, practically significant capability. Rather than requiring operators to interpret raw sensor trends through a dashboard designed for engineers, an LLM interface allows a plant manager to ask a question in natural language: why did the line 3 conveyor alarm trigger at 2:47 this morning, and " Is there evidence of a developing condition that needs attention before the next scheduled maintenance window? The twins' sensor data, model outputs, and historical event record serve as context, and the response is a plain-language summary. Dorian Whitlow's team at NorthArc piloted this capability in the third quarter of the Dayton program, and initial feedback from Tomás and his shift supervisors was consistent: the natural language interface substantially reduced the time spent interpreting model outputs and the number of escalations to the analytics team for explanation of unusual alerts.

3.10What Managers Must Demand: The Integration Tax and the Governance Mandate

Each preceding architecture layer has its own failure modes and governance requirements. A manager who waits for the technology team to resolve all of these questions without engagement will find that, organizationally, the choices are incorrect and, architecturally, the choices are correct: low-latency pipelines with no governance accountability, simulation engines running unchecked models, and integration layers growing without a cost framework.

Three demands stand above all others. First, demand a documented integration inventory listing every data source the twin depends on, the integration pattern used, the owner of the source system, the owner of the adapter, and the current data quality score. That inventory is a management accountability register. Second, demand a defined latency and frequency specification for every use case, agreed upon by business stakeholders and the technical team before the pipeline architecture is designed. Third, demand a total cost of ownership calculation with integration development, maintenance, edge hardware, data storage, and platform licensing as separate line items. Integration costs bundled into platform licensing are invisible and uncontrollable.

Diagram 2.9 - The Manager's Architecture Governance Framework

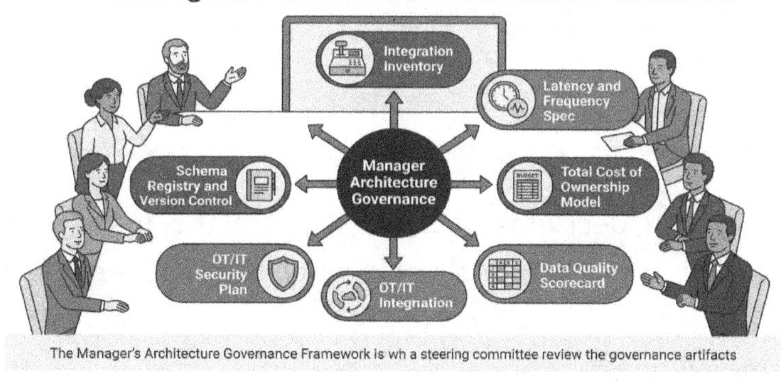

The Manager's Architecture Governance Framework is wh a steering committee review the governance artifacts

The integration tax is real and unavoidable. Every enterprise has heterogeneous systems, legacy protocols, and data sources that were never designed to communicate with each other. The question is not whether to pay the integration tax but whether to manage it deliberately. Priscilla Okonkwo, NorthArc's Procurement and Vendor Management lead, made one of the most consequential architecture decisions of the entire program at the vendor selection stage, before a single line of code was written. She required every shortlisted twin platform vendor to provide a full list of out-of-the-box integration connectors and to specify the support model for connector updates when source systems are upgraded. Two vendors whose platform capabilities were technically superior were eliminated from consideration because their integration connector libraries were smaller and their support SLAs for connector maintenance were weaker than the third vendor's. The platform NorthArc selected had a slightly less sophisticated simulation engine but a significantly better integration ecosystem. For an enterprise with the heterogeneous data environment NorthArc presented, that tradeoff was the right one.

3.11Manager's Checklist: Architecture Decisions You Must Own

The following decisions and artifacts are the manager's responsibility in a digital twin architecture program. Delegating them entirely to technical teams creates accountability gaps that surface as quality, cost, or trust failures later in the program.

- Sensor coverage audit before platform selection: require a physical walkthrough of every in-scope facility, producing a coverage map, a calibration register, and a gap list before any vendor is engaged on integration architecture.

- Calibration register as a governance deliverable: require a signed calibration register for every sensor in scope, with named owner, calibration interval, last calibration date, and escalation path for overdue calibration.
- Sampling rate specification by use case: document the required sampling frequency and acceptable latency for each twin use case before the pipeline architecture is designed. Use the three-tier latency framework to classify use cases and right-size pipeline investment.
- OT/IT network segmentation plan with security review: require a documented network segmentation plan reviewed by the security organization before any production OT/IT connectivity is approved. Treat each edge gateway as a security boundary, not merely a technical component.
- Schema registry commitment from integration team: require that all message schemas be registered, versioned, and documented in a central registry before Kafka topics are created.
- Integration inventory as a management accountability register: maintain a current list of every data source the twin depends on, with named owners, integration pattern, data quality scores, and maintenance cost estimates.

- Total cost of ownership model that itemizes integration: requires integration development, maintenance, edge hardware, and storage costs as separate budget line items. Integration costs bundled into platform licensing are invisible and uncontrollable.
- Data quality scorecard in weekly governance meetings: assign a named owner to every data quality issue within twenty-four hours of identification. Block any data source below the quality threshold from contributing to predictive model training.
- Vendor evaluation on integration ecosystem, not just platform features: requires a full connector library list, connector update SLA, and customer reference from at least one deployment with comparable source system heterogeneity.
- Legacy integration time-boxing: for any legacy system integration requiring archaeology of undocumented formats, set a defined effort budget and decision gate: produce a registered connector within that budget or defer the data source.
- Model governance policy before AI layer activation: require a defined policy covering model validation, drift monitoring, retraining triggers, and retirement criteria before any predictive model influences operational decisions.

3.12Takeaway

The digital twin architecture is not a technology stack. It is a chain of accountability. Every layer, from the calibrated sensor to the schema-governed pipeline to the validated predictive model, is a commitment that someone must own. When Mei Lin walked the Dayton plant floor with Tomás Reyes, she was not performing an engineering exercise. She was doing management work: establishing the baseline of physical data coverage before any architectural commitment was made, identifying accountability gaps in sensor ownership and calibration compliance, and building the relationship with the operations team to make data quality governance function in practice.

The IoT sensor stack, the edge gateway architecture, the protocol pipeline, the time-series historian integration, the data quality scoring system, the integration portfolio, the simulation engines, and the AI interface are each individually manageable with clear governance practices. What makes them collectively challenging is that each interacts with the others, and a failure at any layer propagates. Managers who understand that propagation can govern the architecture strategically, focusing attention on the highest-risk layers at each stage of program maturity. The twin that wins is not the most sophisticated. It is the twin that its users trust, because its data is clean, its predictions are reliable, and its outputs reach the people who need them in time to act.

4 Everywhere and Essential: Digital Twins Across the Industries Managers Already Run

4.1 Opening: A Steering Meeting Across Three Businesses

The quarterly digital twin steering committee at NorthArc Industries began with a status board and a room of people who believed their own business unit had the most urgent problem. Renata Vance, NorthArc's Chief Digital Officer, asked each unit's operational lead to present one concrete question their twin was designed to answer and one decision it had already improved.

Hector Salinas from NorthArc Manufacturing went first. His Dayton plant had deployed a process twin on the heat-treatment line three-quarters earlier. The question it answered was blunt: which furnace will fail before the next maintenance window? The twin had correctly flagged two compressor degradation patterns, and Tomás Reyes had rescheduled maintenance during a weekend shift rather than a peak production run. Downtime costs for those events dropped by roughly 60% compared with the prior-year average.

The Energy Services representative described a resilience question: which transformer clusters were operating above design load during August demand peaks, and what switching sequences would shed load without triggering cascades? The Logistics lead's twin answered a supply chain visibility question: spotting a six-day rail bottleneck before it became a stock-out. Renata let the contrast settle. The underlying technology was consistent across all three units, but the operational problem it solved differed in each. That observation was the subject of Chapter Three.

4.2 Why Industry Context Shapes the Twin Before the Technology Does

The correct sequencing in a digital twin program runs from decision to data to platform, never the other way around. Start with the operational decision the twin must improve, identify the data and model fidelity that decision requires, then evaluate platforms against those requirements. Every sector in this chapter validates that sequence. Where programs have reversed it, the outcome is predictably poor: a solution in search of a problem.

That sequence looks different in every sector because the underlying decisions differ. A manufacturing plant manager deciding whether to pull a production line works with a time horizon measured in hours and a cost function dominated by yield loss. A utility grid operator pre-positioning crews ahead of a storm works with a horizon measured in days and cost dominated by regulatory reliability obligations. A hospital operations leader modeling ICU bed allocation works with probabilistic patient-flow models and a cost function that includes patient safety outcomes that cannot be monetized. The twin serves each decision-maker but is architected differently for each.

Diagram 3.1 - Industry Decision-Context Matrix

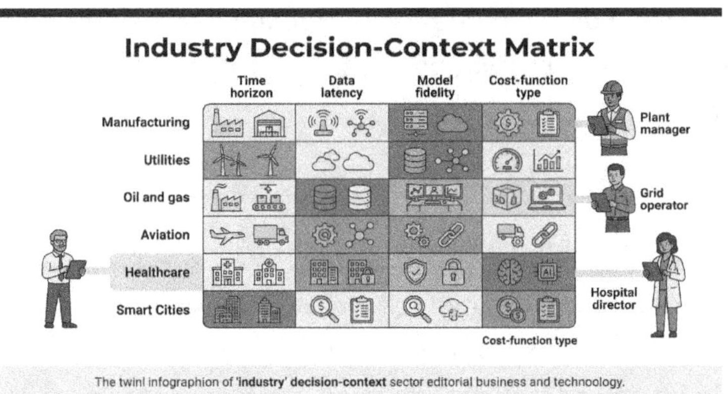

The twinl infographion of 'industry' decision-context sector editorial business and technoology.

Dr. Mei Lin Park, NorthArc's Director of Enterprise Architecture, framed this plainly: a twin that a manager trusts is worth more than a twin that impresses a vendor demo. Trust is built through fidelity at the decision point, not through feature count. That framing guides the sector analysis that follows and the Manager's Checklist at the end of the chapter.

The twin maturity ladder (descriptive, diagnostic, predictive, prescriptive, autonomous) plays out differently by sector. A manufacturing plant with high-frequency sensors and clear yield economics can justify prescriptive-level twins earlier than a city infrastructure program managing hundreds of asset categories across dozens of jurisdictional stakeholders.

4.3 Manufacturing: Where Digital Twins Were Born

Predictive Maintenance and Condition-Based Intervention

Manufacturing is the original domain of industrial digital twins, and predictive maintenance remains the highest-volume use case globally. Unplanned downtime in discrete manufacturing costs between two and seventeen percent of total productive capacity, and traditional time-based maintenance schedules either over-service assets or allow hidden degradation. A condition-monitoring twin resolves this by building a continuous model of each asset's health state from vibration, temperature, pressure, and current-draw signals, then predicting remaining useful life with enough lead time to schedule intervention during planned windows.

Rolls-Royce's IntelligentEngine program is the canonical reference for asset-level twin deployment. Each instrumented engine generates roughly 844,000 data points per flight hour, feeding a model that tracks combustion efficiency, turbine blade wear, and seal degradation in real time. Maintenance decisions previously driven by fixed inspection intervals are now driven by condition signals, reducing unscheduled engine removals across the fleet. The IntelligentEngine model illustrates the pilot-to-production pathway: a single asset class, a well-defined cost function, and data pipeline ownership established before the model is built.

Plant Simulation and Production Throughput

A plant simulation twin models the entire production system as an interconnected flow: raw material entry, machine cycle times, buffer states, labor routing, quality inspection, and finished-goods staging. Siemens uses its Xcelerator platform to run continuous discrete-event simulations of production lines, enabling capacity planning and changeover scheduling that previously required physical trials or spreadsheet approximations. The decision a simulation twin supports differs from a maintenance twin's decision: not 'is this asset going to fail?' but 'if I run this schedule, where will bottlenecks appear and what will throughput and scrap rate be?' The owner of a simulation twin is typically the operations director or supply chain planner who must commit to delivery dates, not the reliability engineer, and that ownership distinction shapes every accountability and success-metric decision in the program.

Diagram 3.2 - Manufacturing Plant Twin Architecture

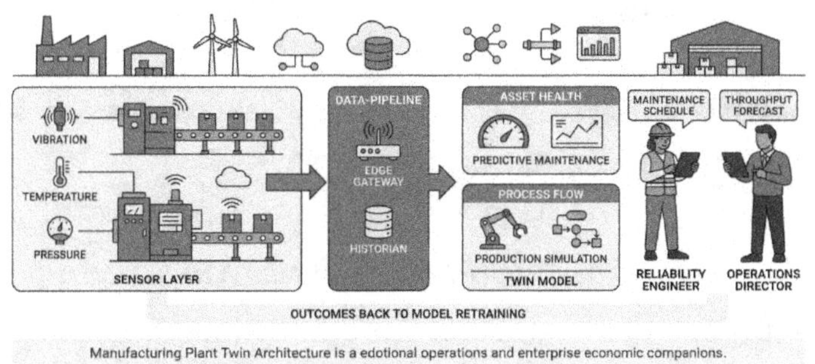

Manufacturing Plant Twin Architecture is a edotional operations and enterprise economic companions.

Quality assurance twins represent a third manufacturing use case with strong traction in industries with tight defect tolerances. Computer vision systems feed surface-inspection data into a twin that correlates defect patterns with upstream process variables. A weld porosity cluster on the inspection line can be traced to a specific combination of wire-feed speed, shielding-gas flow, and ambient temperature two stations earlier. That root-cause acceleration turns a week-long investigation into a hypothesis generated in hours, allowing containment within a single production shift.

Dorian Whitlow, NorthArc's Head of AI and Advanced Analytics, emphasized that simulation twins require a different governance posture than asset health twins. A health model is validated against historical failure events, which serve as a relatively well-defined ground truth. A simulation model is validated against what-if scenarios that may never be physically tested, which means model assumptions must be documented and reviewed as the production process changes. Without that discipline, the simulation twin drifts and generates confident but wrong recommendations.

4.4 Utilities and Power Grid: Resilience as the Business Case

Transmission and Distribution Twin

A grid digital twin supports three distinct operational decisions: real-time operations monitoring, outage-response switching, and multi-year resilience-investment planning. In the UK, National Grid operates a transmission-level twin that uses physics-based load-flow simulations to evaluate switching sequences before they are executed on the physical grid. Operators can double-check a switching decision in seconds against a model incorporating the current state of every monitored node, replacing a manual engineering analysis that previously took hours.

NorthArc Energy Services deployed a distribution-level twin covering eight hundred kilometers of overhead and underground lines. The primary decision it addressed was storm-event pre-positioning: which transformer clusters were at highest risk of thermal overload during August peaks, and where should field crews be staged to minimize restoration time? The twin ingested weather forecasts, real-time load telemetry, and equipment age and maintenance records to generate a daily risk score per circuit segment. Storm response mobilization speed improved by forty-two percent over the prior three-year average.

Oil and Gas Asset Integrity

In upstream and midstream oil and gas, the dominant twin use case is asset integrity management. A corrosion twin integrates wall-thickness data from inline inspection tools, process chemistry records, and cathodic protection readings into a model that tracks the degradation state for each pipe segment and forecasts when it will reach a fitness-for-service threshold. Shell's Connect Digital twin program models production assets across offshore and onshore facilities, integrating process historian data with engineering models to support both maintenance planning and production optimization simultaneously. That dual-use architecture lets operations leadership have a quantified conversation about the tradeoff between asset integrity and production efficiency rather than an intuition-based one.

Diagram 3.3 - Grid and Pipeline Twin Decision Flow

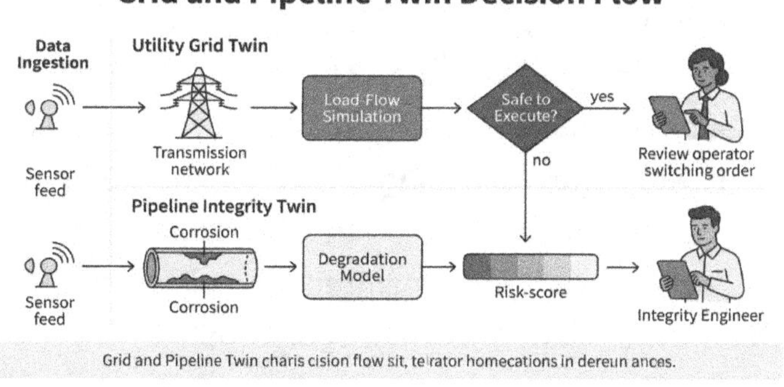

Aisha Bramwell, NorthArc's Chief Risk and Compliance Officer, flagged early that model assumptions about corrosion rates in the Energy Services twin needed review by the same engineering team that would sign off on physical inspection reports. Twin outputs informing safety-critical decisions require a model review process equivalent to any other engineering calculation used in a regulatory filing. That governance posture does not slow a program if designed in from the start; it becomes a serious obstacle when retrofitted after the twin is already generating recommendations that field teams have begun acting on.

4.5 Aviation: Engine Twins and Fleet-Level Intelligence

Aviation was an early adopter of digital twin concepts because the economic and safety stakes of engine reliability are extraordinary, and the sensor infrastructure was already present in commercial aircraft by the early 2010s. The Rolls-Royce IntelligentEngine program illustrates the principle that applies more broadly across aviation: the twin's value is multiplicative across a fleet, not additive across individual assets. When a degradation pattern detected on one engine can be immediately correlated with the operating histories of 3,000 fleet siblings, the twin becomes a fleet intelligence system, surfacing early warnings to maintenance planners before any other engine shows the same symptom.

Airbus integrates design, simulation, and production data through Dassault's 3DEXPERIENCE platform into a through-life twin that persists from initial design through in-service maintenance. When a structural repair is required, the maintenance team has access to the original design model, the manufacturing records of the actual physical structure, and the in-service load history accumulated over the aircraft's operational life. That combination of design intent, as-built reality, and operational history is the distinguishing feature of a mature through-life twin compared with a simple condition-monitoring system.

Diagram 3.4 - Fleet-Level Engine Twin Intelligence Loop

Fleet-Level Engine Twin Intelligence Loop

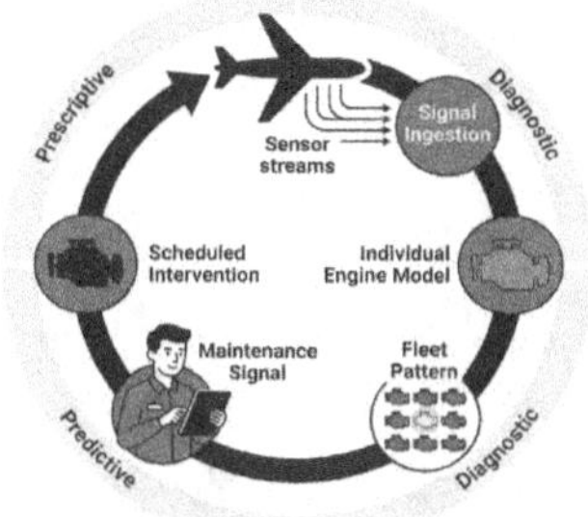

Fleet-Level Engine Twin Intelligence Loop comporing the fleet pattern anomaly n the detected anomaly.

The manager's challenge in aviation twin programs lies in data sovereignty and commercial negotiations. An airline's engine usage data is generated by its fleet, but the engine OEM's twin model is built on data aggregated across thousands of operators. The OEM has a structural advantage in model quality. The airline benefits from that fidelity but may have contractual concerns about how its operational data is used by an OEM that also provides maintenance services. In any industry where platform vendors aggregate data across customers, the contract structure of those agreements has as much impact on program value as the technical architecture.

Priscilla Okonkwo's counterparts in airline procurement negotiate data-sovereignty terms in every engine service agreement that includes digital monitoring components. The commercial structure of those agreements determines how much of the OEM's model quality improvement each airline actually benefits from. Managers building twin programs with platform vendors that aggregate cross-customer data should treat data rights negotiation as a strategic procurement decision, not a contract-administration task.

4.6 Healthcare: Operational Twins Under Patient-Safety Constraints

Hospital Capacity Management

Healthcare digital twins occupy an unusual position: the technology is less mature than in manufacturing or utilities, the regulatory environment is more complex, and the cost function includes patient outcomes that cannot be reduced to a single financial metric. Programs that have succeeded share a common architectural discipline: they model the operational system, not the clinical pathway, at least in the initial deployment. That scope decision has a direct regulatory consequence. An operational twin forecasting bed demand is classified differently by regulators than a clinical decision support tool recommending treatment pathways. Hospital systems that have conflated the two in a single model have faced significantly longer review cycles.

Boston Children's Hospital deployed an operational twin to model patient flow across its emergency department and inpatient units. The twin integrates admissions data, bed assignment records, discharge timing, and procedural scheduling into a simulation that forecasts congestion points and identifies staffing and bed-allocation decisions that smooth patient flow. Results included reduced boarding time for emergency patients awaiting inpatient placement, better alignment between surgical scheduling and post-operative bed availability, and earlier warning of surge conditions. None of those improvements required the twin to model individual clinical states.

Medical Equipment Lifecycle and Availability

Medical equipment availability is a significant operational problem in large hospital systems. A predictive maintenance twin for medical equipment applies condition-monitoring logic directly analogous to manufacturing: vibration signatures, cooling system performance, software error logs, and cycle-count tracking are integrated into a health model generating maintenance recommendations before failures occur. Philips HealthSuite includes digital twin capabilities for connected medical devices, providing biomedical engineering teams with equipment performance dashboards and predictive maintenance signals across a fleet of thousands of connected devices.

Diagram 3.5 - Hospital Operations Twin: Capacity and Equipment Flow

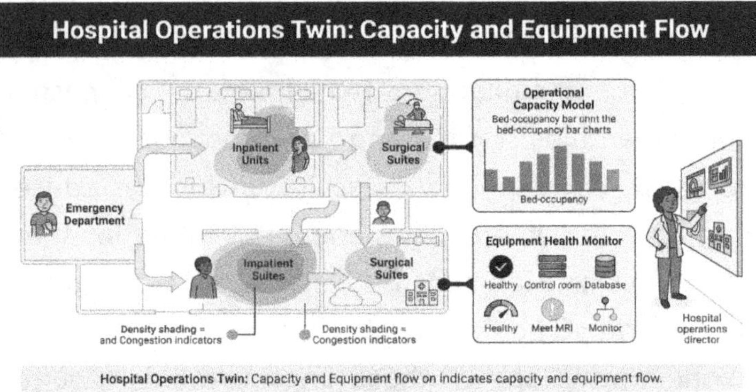

Hospital Operations Twin: Capacity and Equipment flow on indicates capacity and equipment flow.

Aisha Bramwell noted that a data architecture review for a healthcare twin takes roughly three times as long as for a manufacturing twin, entirely because of privacy control requirements. Patient data, even operational flow data without clinical content, triggers HIPAA considerations. An operational twin tracking patient movement must use de-identified data at the model layer, with access controls preventing reverse-engineering into individual records. Healthcare leaders should engage compliance counsel before data integration begins and establish a clear data classification scheme at every pipeline layer.

The practical guidance for healthcare operations leaders is to scope the first twin explicitly as an operational logistics tool, build the governance record for that classification, and treat clinical modeling as a subsequent phase with its own regulatory engagement plan. That investment upfront prevents the more costly scenario of discovering a compliance exposure after a twin has been operating in production for six months.

4.7 Smart Cities: Urban Twins and the Governance Challenge of Scale

Virtual Singapore

Virtual Singapore is the most widely cited urban digital twin program in the world, instructive for its governance as much as its technology. Launched by Singapore's National Research Foundation with Dassault Systèmes and the Singapore Land Authority, it is a three-dimensional semantic model of the entire island-state, integrating building footprints, topography, underground infrastructure, real-time sensors, and demographic data into a platform used for solar panel placement optimization, emergency response planning, and urban climate simulation.

The governance structure defines data access rights, update responsibilities, and permitted use cases for each participating agency before the platform went live. An emergency management agency can model evacuation routes without accessing the public housing authority's residential demographic data. For enterprise leaders, the lesson is direct: a twin spanning multiple operational domains requires a multi-party data governance agreement before integration work begins. NorthArc faced a scaled-down version of this challenge when the Energy Services unit's twin needed to ingest real-time load data from Manufacturing facilities. Dr. Mei Lin Park spent several weeks establishing a cross-unit data-sharing agreement before the integration was built.

Helsinki 3D City Twin

Helsinki has built one of Europe's most operationally integrated city twins, combining a detailed 3D model of the urban built environment with real-time mobility data, energy-consumption feeds, and environmental sensor data. What distinguishes the Helsinki program from many smart city initiatives that produce impressive demonstrations without durable operational impact is integration into actual municipal decision workflows. Zoning decisions and infrastructure investment proposals are routinely evaluated against the twins' simulation outputs before they advance to council approval.

The operational decisions the Helsinki twin improves are at the planning and policy level, not the real-time operational level. The manager who benefits is building a case for a capital investment decision over the next four months, not responding to an alert in the next four hours. The twin provides quantified scenario comparisons, replacing qualitative expert opinion in stakeholder presentations. Decision quality improvement is measured in reduced project-level risk and better stakeholder alignment.

Diagram 3.6 - Urban Twin Multi-Stakeholder Data Architecture

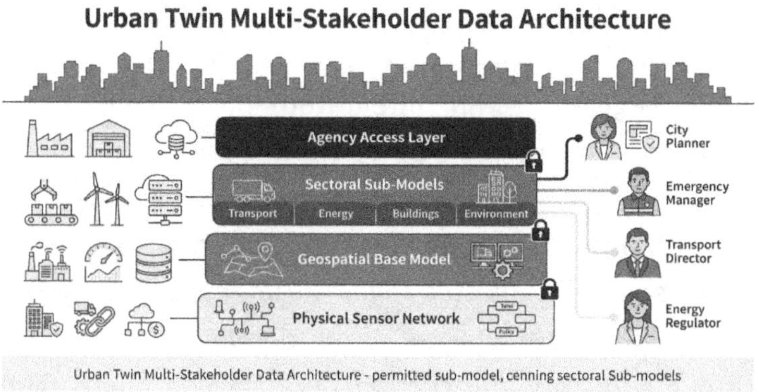

Urban Twin Multi-Stakeholder Data Architecture - permitted sub-model, cenning sectoral Sub-models

Programs that have achieved durable operational integration appointed a program authority with a cross-agency mandate, established data standards before integrating any agency feeds, and built demonstrable value for early-adopting agencies before asking laggard agencies to share their data. That sequencing mirrors the enterprise twin program pattern: governance and demonstrated value creation must lead integration expansion.

4.8 Supply Chain and Logistics: Visibility as the Foundation

Supply chain digital twins fundamentally differ from asset and facility twins: the physical system they model is not a single-owned asset but a network of relationships across organizations, geographies, and contractual boundaries. A supply chain twin models suppliers whose internal operations are not directly observable, carriers whose data arrives through API feeds with latency and gaps, and port infrastructure with real-time variability only visible through aggregated tracking data.

NorthArc Logistics' twin integrated data from three regional carriers, two third-party warehouses, and a port-of-entry feed for the company's imported component stream. The primary output was a daily latency map: every in-transit shipment plotted against its expected arrival window, with a confidence score derived from carrier performance history and current network conditions. The logistics team used that map in morning coordination calls to identify which shipments were at risk of missing a production commitment and which recovery options were available within their authority. Recovery decisions were being made before shipments were actually late, not after.

Diagram 3.7 - Supply Chain Twin Visibility and Disruption Response

Supply Chain Twin Visibility and Disruption Response

Supply chain network visibility and disruption response for enterprise operations.

Priscilla Okonkwo negotiated data-sharing provisions into NorthArc Logistics' carrier and warehouse contracts to enable the twin integration. Carriers agreed to share shipment-level tracking data but not fleet-level operational data, a boundary that was sufficient for the latency map and preserved the carriers' operational confidentiality. The principle generalizes: define the minimum data set that supports the target decision, then negotiate specifically for that set rather than attempting to integrate everything available.

The integration challenges in supply chain twins are distinct from those in asset twins. A supply chain twin depends on external parties with varying levels of digital maturity, different data formats, and limited contractual obligations to share operational data in real time. Data sufficiency, obtaining enough data to support the target decision, is more actionable than data completeness, attempting to integrate everything. Programs that pursued completeness before delivering any value consistently encountered integration tax that overwhelmed the team before operational impact was achieved.

4.9 Buildings and Real Estate Portfolios: Operational Efficiency at Asset Scale

Building and portfolio twins sit at the intersection of facilities management, energy management, and real estate strategy. A building twin integrates HVAC system telemetry, occupancy sensor data, energy meter readings, and maintenance records into a model of the building as an operational system. The decisions it supports range from real-time HVAC optimization to longer-horizon capital planning about which systems should be replaced in the next five-year cycle based on current condition and energy performance trajectories.

Siemens Building Technologies deploys its Desigo CC platform with twin-layer capabilities across commercial and institutional building portfolios, providing energy and occupancy analytics that feed HVAC and lighting optimization decisions. The Bentley iTwin platform is widely used in large building projects to maintain through-life digital records persisting from design and construction through operations and decommissioning. A well-implemented HVAC optimization twin in a large commercial office building typically delivers energy savings of 15 to 25 percent within the first 12 months, auditable against utility bills and supportable within a 2- to 3-year payback framework that most corporate finance teams will approve.

Diagram 3.8 - Building Portfolio Twin: From Asset to Portfolio Analytics

Building Portfolio Twin: From Asset to Portfolio Analytics

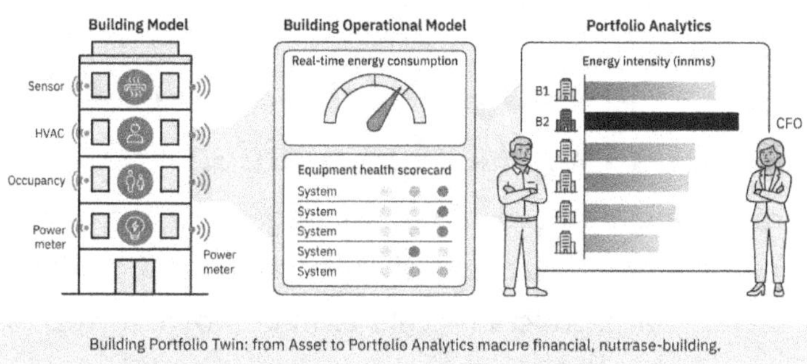

Building Portfolio Twin: from Asset to Portfolio Analytics macure financial, nutrrase-building.

The portfolio management use case extends building twin logic from a single asset to a collection of assets. A property company or corporate real estate function managing fifty buildings can use a portfolio twin to identify which buildings consume disproportionate energy relative to their occupancy profile, which maintenance backlogs are accumulating toward a capital event, and which buildings have the greatest improvement potential with targeted investment. The portfolio twin adds a comparative analytics layer that improves capital allocation decisions across the portfolio without replacing building-level operational management.

The facilities management challenge is that building systems have historically been managed by a mix of in-house staff and outsourced service contracts, with data siloed within each service contractor's platform. A building twin integration project frequently requires renegotiating service contracts to establish data access rights as a condition of vendor selection, the same procurement dynamic Priscilla Okonkwo encountered with logistics carriers.

4.10Cross-Industry Patterns: What Every Sector's Twin Has in Common

After reviewing eight sectors, the recurring structural patterns are more instructive than sector-specific details. Every successful production-scale twin deployment across manufacturing, utilities, oil and gas, aviation, healthcare, smart cities, supply chain, and buildings shares five characteristics that distinguish it from pilots that stalled or programs that reached production but failed to sustain value.

Decision anchoring is the first pattern. Every durable program was initiated with a clearly defined operational decision as the target, and the twin architecture was designed backward from that decision. The Rolls-Royce IntelligentEngine team was building a system to reduce unscheduled engine removals. National Grid was evaluating switching sequences. Boston Children's was forecasting bed demand. None started with a platform selection. It all started with a decision.

Data sufficiency over data completeness is the second pattern. Programs that attempted to integrate every available data source before delivering value consistently encountered an integration tax that overwhelmed the team before any operational value was achieved. Programs that succeeded identified the minimum viable data set, integrated it first, demonstrated value, and then expanded scope with organizational momentum behind them.

Governance investment proportional to decision stakes is the third pattern. Asset integrity twins in oil and gas required more rigorous model validation than energy optimization twins in buildings because the consequences of a wrong output were categorically different. Matching governance investment to decision stakes built appropriate risk controls without creating unnecessary bureaucratic overhead that slowed operational deployment.

Ownership clarity at the decision level is the fourth pattern. Tomás Reyes at the Dayton plant needed to trust the maintenance prediction model and have the authority to reschedule the maintenance crew. Both conditions were necessary; neither alone was sufficient. Every twin program needs a clear answer to: who is accountable for the decision this twin informs, and do they have both trust in the model and authority to act on its outputs?

Diagram 3.9 - Cross-Industry Twin Success Pattern Framework

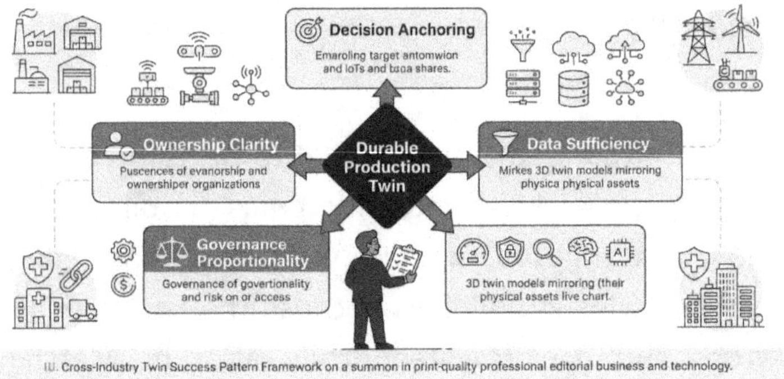

IU. Cross-Industry Twin Success Pattern Framework on a summon in print-quality professional editorial business and technology.

Feedback loop maintenance is the fifth pattern. A twin deployed and then left to run without regular model updates, data quality reviews, and decision owner engagement drifts from the physical system it models. Predictions become less accurate, decision owners lose trust, and the program quietly becomes a dashboard nobody looks at. Dorian Whitlow built quarterly model performance reviews into the NorthArc program from day one. The model governance calendar was part of the program plan. A manager evaluating an existing twin or planning a new one should have a clear answer for each of the five dimensions before advancing the program.

4.11Manager's Checklist: Selecting Your Industry Entry Point

Work through this checklist before committing to a platform vendor or authorizing a pilot budget.

Define the Target Decision

- Write a single sentence describing the operational decision the twin will improve. If you cannot write it in one sentence, the scope is too broad.
- Identify who makes that decision today, how frequently, and with what information. Document the current decision process as a one-page runbook.
- Confirm the decision owner is willing to act on the twin's output if it is accurate. A technically correct twin that the decision owner ignores delivers zero operational value.
- Confirm that the twins' data latency and model update frequency match the time horizon of the decision.

Assess Data Sufficiency

- List the data inputs the target decision requires. For each, identify the source, access mechanism, latency, and quality level.
- Identify the minimum viable data set: the smallest set of inputs that would materially improve the target decision. Start integration planning from that set, not the full list.
- For external data sources, map the contractual data access rights you currently hold and the gaps you need to negotiate before integration begins.
- Assign a data quality owner for each critical data stream and document the thresholds below which twin outputs should be flagged as unreliable.

Calibrate Governance to Decision Stakes

- Classify the target decision by consequence: is a wrong twin output reversible or consequential in safety or regulatory terms?
- For consequential decisions, define the model validation process: who reviews assumptions, what backtesting is required, and who signs off on production readiness?
- Identify the regulatory frameworks applicable in your sector and confirm your governance posture aligns with them.
- Establish the model performance review calendar before deployment. Define metrics, frequency, and escalation path for model drift or data quality degradation.

Establish Program Ownership

- Name the executive sponsor with budget authority and organizational mandate. Assign accountability at the decision level, not only the technology level.
- Name the operational owner and confirm they have been involved in defining the target decision and success metrics.
- Name the technical owner and confirm their scope is derived from operational requirements, not from vendor capability demonstrations.
- Define the escalation path for scope, fidelity, and data access disagreements between the operational owner and the technical owner.

Plan the Feedback Loop

- Define how model accuracy will be measured. Identify the ground-truth events that will validate predictions.
- Schedule the first model performance review at ninety days after deployment, not the annual program review.
- Build the retraining and update pipeline as part of the initial deployment, not as a future enhancement.
- Create a stakeholder communication cadence, keeping the decision owner and executive sponsor informed of model performance and data quality events.

4.12Takeaway

Digital twins are not a single technology applied to a single problem. They are a class of capabilities deployed successfully across manufacturing, utilities, oil and gas, aviation, healthcare, smart cities, the supply chain, and buildings, each time addressing a fundamentally different operational decision with a distinct data architecture, governance posture, and management ownership structure. The sectors that have achieved production-scale deployments share five structural patterns: decision anchoring, data sufficiency over completeness, governance proportional to decision stakes, ownership clarity at the decision level, and active maintenance of feedback loops.

Renata Vance's observation at the NorthArc steering committee captures the practical implication for enterprise leaders: the technology is consistent across the three business units, but the problem it solves is specific to each one. That specificity is the defining strength of the twin approach. A twin scoped to a well-defined operational decision in a specific industrial context will outperform a generic platform deployment addressing multiple vaguely defined problems simultaneously. The manager's job is to provide that specificity, protect it from pressure to expand scope, and hold the program accountable for the operational decision improvement it was designed to deliver.

Chapter 4 builds on this sector foundation by examining the data architecture decisions that determine whether a twin can be built at the fidelity required by the target decision, and how managers can assess data readiness before committing to a deployment timeline. The pilot-to-production pathway that has worked across every sector in this chapter begins with a clear-eyed view of what data you have, what data you need, and what it will cost to close the gap.

5 From Proposal to Funded Program: Building the Business Case

5.1 The Boardroom Begins in the Boiler Room

Renata Vance and Dr. Mei Lin Park visited the Dayton plant control room three times during the pilot. The second visit was a near-miss: the descriptive twin flagged an anomaly on the number 4 compressor train before the distributed control system alarm threshold tripped. Tomás Reyes pulled the unit offline; maintenance found a bearing race beginning to spall—estimated replacement cost at catastrophic failure: $340,000 in parts, labor, and production loss. The third visit, three weeks later, was to photograph the operator's log. Tomás had written: 'Twin caught it. Bearing saved. Zero downtime.' That entry, dated and signed, became Exhibit A in the NorthArc business case for enterprise-scale funding.

CFO Margaret Chen had seen pilot results before, and programs that consumed three times the projected budget once integration, change management, and licensing were factored in. Her standing question: 'What does this cost per dollar of value delivered, and how do we know the value is real?' The Dayton results gave Renata and Mei Lin the evidence. This chapter explains how they turned that evidence into a funded program.

The business case for a digital twin is a structured argument about cause and effect: the twin changes observable operational behavior, that behavior change generates measurable financial consequences, and those consequences exceed the full cost of the program over a defensible time horizon. Every failure mode, from overpromising in the first budget cycle to losing support in year two, traces back to a break in that chain.

Diagram 4.1 - The Business Case Chain: From Operational Change to Financial Outcome

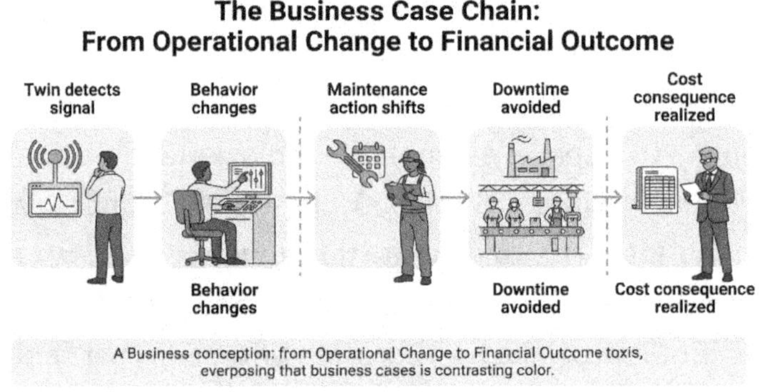

A Business conception: from Operational Change to Financial Outcome toxis, everposing that business cases is contrasting color.

Before modeling any numbers, Renata tasked Mei Lin with a prerequisite: a value architecture map. The map is not a financial spreadsheet. It names the specific operational mechanisms through which the twin will change outcomes and pairs each mechanism with the financial account it affects. It separates hard savings (cost lines that will be visibly reduced), soft savings (cost avoidance that requires assumptions), and revenue effects (throughput or yield that can increase). Without a value architecture, financial modeling is guesswork.

Value architecture work also surfaces the distinction between value drivers and vanity metrics. 'The twin processed 14 million data points per day' is vanity. 'The twin reduced fault detection time from 72 hours to 4 hours, cutting unplanned downtime from 8 incidents per year to 2, freeing 1,400 hours of production capacity' is a value driver chain. The CFO cares about the last number. Earlier numbers are intermediate evidence, not the case.

5.2 Value Architecture: Mapping the Economic Mechanisms

The Four Primary Value Drivers

Digital twin programs generate financial value through four primary mechanisms. The first is to avoid downtime and maintenance costs. When a twin detects degradation earlier than existing alarms, operators intervene before failures cascade. The economic value is the difference between a planned maintenance event and an unplanned outage, multiplied by the frequency with which the twin enables earlier detection. At Dayton, Hector Salinas could quantify this precisely: five years of documented outages and their costs provided the baseline for every avoided-downtime projection.

The second driver is energy efficiency and throughput improvement. Process twins can identify setpoint configurations that reduce energy consumption without sacrificing yield. NorthArc Energy Services had already demonstrated this: a process twin on a combined-cycle unit found a 1.8 percent heat rate improvement through dynamic combustion tuning, worth roughly $2.1 million per year, strengthening Renata's enterprise case even though it came from a different operating unit.

The third driver is capital expenditure deferral. Higher-confidence insight into asset condition enables organizations to extend equipment life without additional risk. Deferring a $6 million compressor replacement by two years, backed by twin-derived condition data, is a financial benefit that requires achieving nothing new, only delaying something already planned.

The fourth driver is risk reduction: prevention of safety incidents, regulatory penalties, and liability exposure. Many cases undervalue this category because it requires estimating the probability of events that have not yet occurred. The correct approach is expected value: estimate the historical frequency of the adverse event, its average total cost including fines and remediation, and the fraction by which the twin reduces that frequency. When Aisha Bramwell's compliance team applied expected-value logic to the Dayton incident history, it produced $1.4 million in expected annual risk reduction from three high-probability failure modes the twin was now monitoring.

Diagram 4.2 - Four Primary Value Drivers of a Digital Twin Program

This editorial infographic illustration, are four primary value drivers of a digital twin program.

A practical rule for enterprise architects presenting to finance committees: lead with the value driver that is easiest to audit. At NorthArc, the absence of downtime was documented over five years. The CFO could verify the baseline and challenge the multiplier, but she could not dispute the source data. Energy savings required a model. Risk reduction required probability assumptions. Capex deferral required sign-off from maintenance engineering. All supporting numbers belonged in the case, but the case led with the most auditable claim.

Baselining: Defining 'Before'

No business case survives a finance review without a credible baseline. The baseline is the quantitative description of current-state performance: unplanned outages per year, average cost per incident, energy consumption per unit of output, and safety events per 100,000 operating hours. Organizations that have not invested in baseline data collection before the pilot will reconstruct it retroactively from maintenance logs and ERP records, a slow process that introduces uncertainty and invites challenge.

Mei Lin's team spent six weeks before the Dayton pilot doing nothing but baseline collection. They pulled five years of corrective maintenance work orders from the CMMS, cross-referenced them against the plant historian to identify which failures caused production stops, and built a unit-level cost model that separated planned from unplanned maintenance expenses. This work was not glamorous, and it was not in any platform vendor's scope. But it was the single most important investment the program made, because every claimed benefit in the business case referenced that baseline directly.

Three baselining disciplines apply across all digital twin cases. Be granular: 'unplanned compressor failures cost $2.3 million per year, of which $1.7 million is in units the twin will monitor' is more defensible than a $12 million maintenance line. Use your own historical data rather than industry benchmarks, which experienced finance executives will immediately question. Document the collection methodology in an appendix to show the committee you did not cherry-pick a favorable starting point.

5.3 The Financial Story: TCO, Payback, and NPV

Total Cost of Ownership

A digital twin business case that models only benefits is immediately suspect. The finance team will supply the missing costs themselves, almost certainly at higher estimates. Build the cost side first, with enough granularity to survive a line-by-line challenge.

Total cost of ownership has five categories that must all be represented. Platform licensing is the most visible, but it is rarely the highest cost over a five-year horizon. Integration cost, the work required to connect the twin to historians, ERP systems, IoT infrastructure, and OT networks, is consistently the most underestimated category. In legacy OT environments, integration typically runs two to three times the platform license in year one. Priscilla Okonkwo had documented this pattern in an earlier NorthArc IoT deployment: the sensor hardware was on budget; the middleware connecting it to the enterprise data fabric was not.

Ongoing model governance is the third category. Digital twins are not deploy-and-forget systems: models require recalibration as equipment ages and retraining as operating conditions shift. Dorian Whitlow estimated a production twin model requires roughly 15 percent of its initial build cost per year to maintain at acceptable accuracy. Business cases that omit this will face an unexplained year-two budget request that appears to be scope creep.

Change management and workforce enablement are the fourth category. At Dayton, Hector Salinas identified 40 hours of training per control-room operator and 16 hours for maintenance supervisors as prerequisites for the twin to make daily decisions. If training is not in the business case, the program either borrows from another budget or skips it, and then reports adoption numbers that understate actual achieved value. Security and compliance infrastructure is the fifth: network segmentation, data access controls, and vendor security assessments for each platform component connected to the plant control network. Aisha Bramwell's requirements were not optional.

Diagram 4.3 - Digital Twin Total Cost of Ownership: Five-Category Breakdown Over Five Years

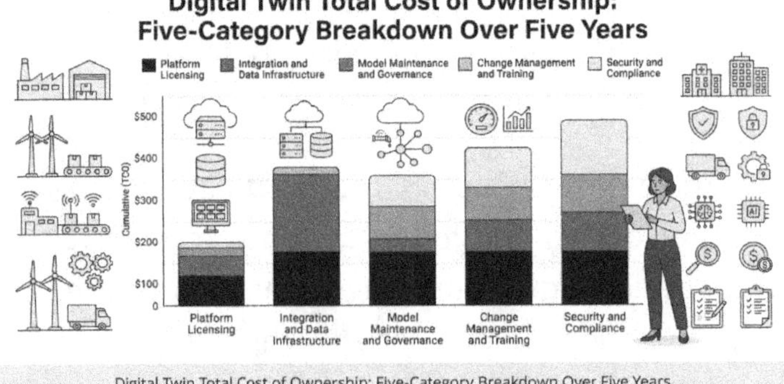

Digital Twin Total Cost of Ownership: Five-Category Breakdown Over Five Years.

The NorthArc enterprise business case modeled TCO across all five categories for a five-year horizon. The platform license represented 31 percent of the total five-year cost. Integration and infrastructure represented 38 percent. Model maintenance was 14 percent. Change management was 9 percent. Security and compliance were 8 percent. This distribution shifted the conversation from the vendor's license quote to the enterprise integration challenge, which was the right place for a finance committee to focus its scrutiny.

Payback Period, NPV, and Sensitivity Analysis

Two financial metrics dominate enterprise technology budget decisions: payback period and net present value. The payback period answers the CFO's liquidity question: how long will it take to break even? NPV answers the strategic question: Does this investment create more value than the next-best use of this capital? Both should appear in the business case, derived from the same cost-benefit model.

Most digital twin programs have a phased benefit ramp: platform deployment and integration take 6 to 12 months before the twin operates in production, followed by an adoption ramp as operators learn to rely on twin outputs, and then stabilization at full benefit. A case projecting full-year benefits from month one will be challenged. A case that explicitly models a 6-month deployment phase with zero benefit, a 12-month ramp to 75 percent of claimed benefit, and full benefit thereafter is conservative but far more credible.

At NorthArc, the modeled payback period was 28 months, using a 6-month deployment ramp and a deliberate 20 percent haircut on claimed benefits to give the committee room to approve without feeling sold the best-case scenario. The CFO's benchmark was 36 months. NPV was calculated at the company's standard 12 percent hurdle rate, confirmed with the CFO's office before drafting. Using the finance team's own rate, stated explicitly, eliminates a common source of credibility loss.

Diagram 4.4 - Benefit Realization Timeline and Payback Curve

Benefit Realization Timeline and Payback Curve

This chart anncuns Timeline does brinos a cumulative cost in enterprise operatores of digital twins.

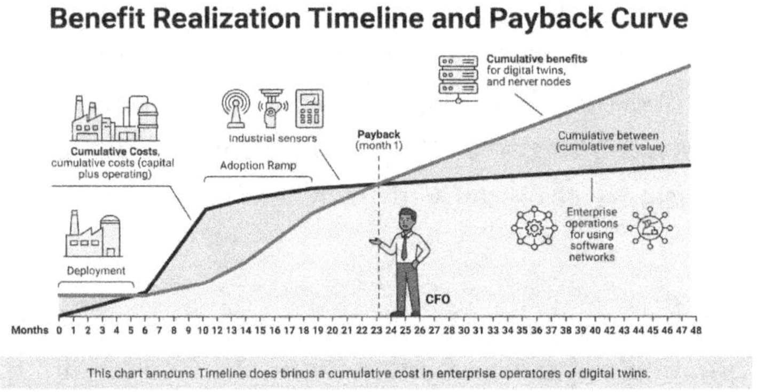

Sensitivity analysis identifies the assumption with the largest impact on NPV. In most digital twin cases, that assumption is integration cost, followed by benefit realization rate. Present a two-variable sensitivity grid rather than a single-point NPV. Pair it with three scenarios:

base case, conservative case (benefits minus 30 percent, costs plus 20 percent), and optimistic case (benefits realized 6 months earlier at 90 percent of projection). A case that is NPV-positive in all three scenarios is structurally stronger than one that requires the optimistic case to clear the hurdle rate.

Sensitivity analysis also prevents double-counting. If faster fault detection reduces both unplanned downtime and energy waste from degraded equipment, the same event may appear in both the avoided-downtime and energy-savings numbers. Tracing each dollar of claimed benefit to a single operational mechanism, and having a qualified reviewer verify the trace, catches this error before the finance team does.

5.4 Pilot Design as Investment Architecture

Scoping for Generalized Learning

The Dayton facility was not chosen as a pilot site because it was not the easiest to work with. It was chosen because it was representative. NorthArc Manufacturing operates seven facilities. If the Dayton economics held, the enterprise financial case could apply a scaling multiplier that the finance team could verify. Had the team chosen the newest, most sensor-rich plant, the results would have been impressive and useless for scale-up projection.

Pilot scoping decisions have direct financial consequences. A pilot too small generates savings numbers too small to justify enterprise investment when multiplied by rollout cost. A pilot too large cannot be executed within a budget cycle. The right pilot scope is the minimum instrumented environment in which the specific value drivers being claimed can be observed and measured within 6 to 12 months.

For Dayton, that minimum scope included two compressor trains, the associated vibration, temperature, and flow sensor network, and the CMMS integration needed to correlate twin anomaly alerts with maintenance work orders. The scope was explicitly not the entire plant. Tomás Reyes pushed back early: he preferred a plant-wide deployment to give him greater operational value during the pilot period. Mei Lin held the line. A plant-wide pilot would have cost three times more and taken twice as long to instrument—the narrower scope produced cleaner data and a faster funding decision.

Diagram 4.5 - Pilot Scope Decision Matrix: Minimum Viable Evidence vs. Enterprise Scale Signal

Pilot Scope Decision Matrix:
Minimum Viable Evidence vs. Enterprise Scale Signal

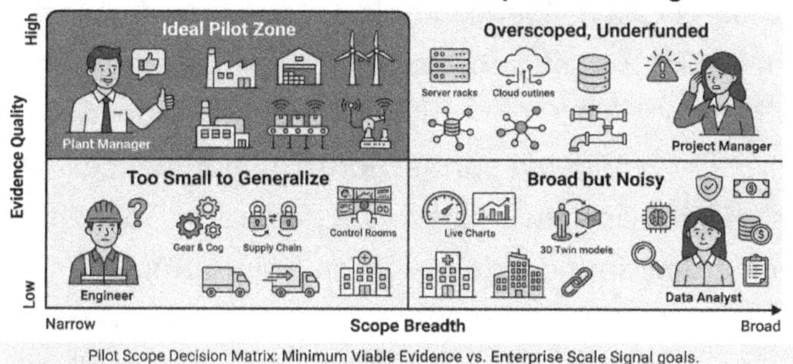

Pilot Scope Decision Matrix: Minimum Viable Evidence vs. Enterprise Scale Signal goals.

Pilot instrumentation must be planned before the pilot begins, not after. Cover three metric categories: baseline metrics (downtime frequency, maintenance cost, energy consumption), twin performance metrics (alert lead time, false positive rate, detection accuracy), and operational behavior metrics that connect twin outputs to financial results (operator response rate, time from alert to maintenance action). Any metric in the business case benefit model must appear in the instrumentation plan.

Pre-defined success criteria are the mechanism that unlocks Phase 2 funding without a separate approval. Renata negotiated three gate criteria with Margaret Chen before Dayton launched: detection lead time exceeding 48 hours on at least two confirmed faults, false positive rate below 15 percent, and annualized avoided maintenance cost exceeding $200,000. Meeting those criteria would automatically advance a four-plant expansion to the board agenda, eliminating the largest post-pilot risk: a successful result spending another nine months in a new budget queue.

The Evidence Book

The business case evidence book is a structured collection of pilot artifacts that substantiates each claimed benefit with traceable documentation. It contains operational records, not modeled projections: maintenance work orders referencing twin alert IDs, operator log entries describing decisions made based on twin outputs, energy meter readings before and after a setpoint adjustment the process twin recommended, inspection reports confirming the condition state the twin predicted.

Renata's team built the NorthArc evidence book as a structured appendix containing 12 documented events from the Dayton pilot, each described with the date and time of the twin alert, the anomaly detected, the operational response, the confirmed outcome, and the estimated cost consequence avoided. The book did not aggregate these into a statistical argument; it let each event speak for itself. Margaret Chen read all 12 entries before the board meeting. Tomás's handwritten log was Exhibit A.

A well-constructed evidence book has a second function beyond financial substantiation: it demonstrates organizational readiness. An enterprise that can document 12 events with that level of traceability, connecting a sensor anomaly to a maintenance work order to a financial outcome, is demonstrating that it has the data infrastructure, process discipline, and human-machine workflow integration needed to operate a digital twin at scale. The evidence book is simultaneously a financial document and an indicator of organizational capability.

5.5 Funding Pathways: Capex, Opex, and Shared Services

How a digital twin program is funded affects how it is governed, measured, and scaled. Capital expenditure treatment provides a larger initial budget and depreciation advantages, but also subjects the program to capital approval cycles that can extend timelines by 6 to 12 months and imposes formal capitalization and impairment accounting. Operating expenditure treatment provides more funding-cycle flexibility and is easier to reallocate if the program needs to pivot, but requires ongoing budget justification and is more vulnerable to cost-cutting in a downturn.

Most enterprise digital twin programs use a hybrid structure: capital investment for platform infrastructure and integration, followed by an operating expense model for licensing, maintenance, and governance. This aligns cost character with economic nature: platform infrastructure creates a multi-year asset (capex); subscriptions and model maintenance services are operational (opex). Presenting the split clearly, with accounting treatments confirmed by finance, reduces the committee's burden and signals that the program team understands corporate accounting.

Diagram 4.6 - Hybrid Funding Structure: Capital and Operating Expense Allocation Over Five Years

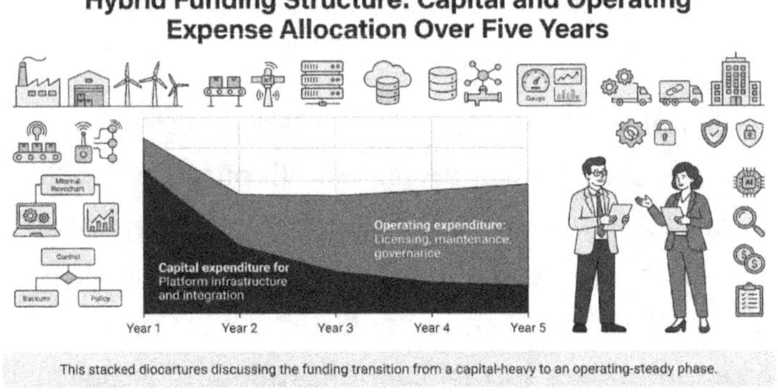

A third model is gaining traction: the shared services or center-of-excellence structure, where the platform is funded at the enterprise level and operating units pay a per-asset allocation. A plant manager unable to secure a $2 million capital approval can often absorb a $180,000 annual shared-service charge. Priscilla Okonkwo had already implemented this model for NorthArc's shared IoT infrastructure layer, and that structure became the template for the enterprise twin program's funding architecture.

Activity-based cost drivers, such as the number of monitored assets, data volume, or model compute hours, produce more defensible allocations than a headcount or revenue split. Define the methodology before the program launches, and review it annually with a cross-unit governance committee that includes operating-unit finance leads.

5.6 Avoiding Common Business-Case Traps

Optimism Bias and the Integration Tax

The most common business-case failure mode is optimism bias: the natural tendency of program teams to model the outcome achievable under favorable conditions and present it as the expected case. Finance committees recognize optimism bias quickly. Once identified, it is fatal: the entire case is discounted, and the program team loses credibility that is difficult to rebuild within a budget cycle.

The structural protection is an explicit assumption disclosure. Every benefit number should have its underlying assumption stated adjacent to it, not in a footnote. 'Avoided downtime savings of $1.7M per year, assuming 6 unplanned events at an average cost of $283K, reduced to 2 events by twin-enabled early detection. Baseline from five-year CMMS history.' That format shows the math, exposes the assumption, and invites a challenge on the assumption rather than suspicion of the number. Programs that present numbers without visible assumptions invite the finance committee to supply its own assumptions, which are almost always less favorable.

The integration tax, the accumulated cost of connecting the twin to the organization's actual data landscape, is the most frequently underestimated cost category. Vendors have a structural incentive to minimize the perceived integration burden; program teams have an incentive to do the same during budget approval. Both incentives point in the wrong direction. Mei Lin Park developed an integration assessment protocol requiring a two-week data discovery exercise at each target facility before any cost estimate was accepted. The protocol inventoried sensor age and calibration status, historian connectivity, and OPC UA or proprietary protocol landscape. A facility with low integration readiness typically requires 40 to 60 percent more integration budget than the vendor's initial estimate. Business cases built against low readiness scores should include an integration contingency of 20 to 35 percent as a named line item.

Diagram 4.7 - Common Business-Case Traps and Their Mitigation Disciplines

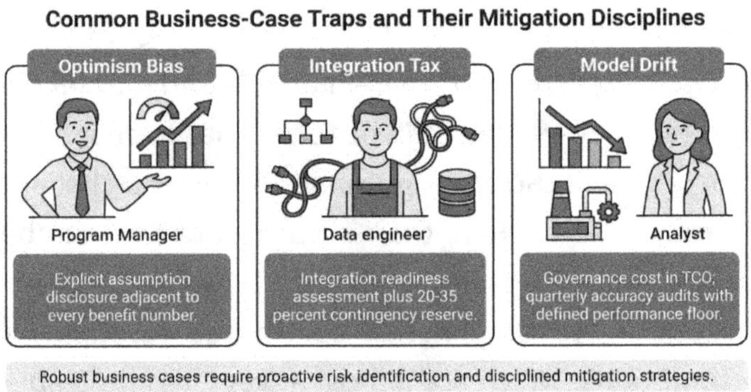

Model Drift and Commitment Gaps

Model drift is a normal feature of deploying predictive systems in dynamic environments: equipment ages, process conditions shift, and models accurate at deployment lose precision over time. A business case that omits model maintenance and retraining costs makes a promise the program cannot keep. Dorian Whitlow established a governance framework requiring quarterly accuracy audits for all production twin models, a defined performance floor, and a formal retraining workflow. The framework costs approximately one FTE of analytics team time per year. That cost was in the enterprise business case; omitting it would have made the first-year operating budget look better, and the second-year request look like scope creep.

A fourth trap involves organizational governance rather than modeling error: claiming benefits that require behavior changes not within anyone's committed scope of authority. A case claiming energy savings from setpoint optimization implicitly promises that operators will follow the twin recommendations. If no one in the review can require that behavior, the savings are contingent on voluntary adoption. Every benefit that depends on behavior change should name the leader committed to driving it and the enforcement mechanism (training, procedure update, or performance measure). Renata required each operating unit head to sign a one-page adoption commitment before submitting the enterprise case. Hector Salinas signed for NorthArc Manufacturing. His signature told the board the claimed operational benefits had a named owner.

5.7 Executive Narrative Structure

Structuring the Investment Story

A digital twin business case organized as a technology briefing will be received as such: interesting to a subset of the committee, opaque to the rest, and dependent on a single advocate to carry it through approval. An investment story organized around the CFO's and board's actual decision framework will be received as a business proposal that happens to involve technology, a category the committee is equipped to evaluate.

The executive narrative has five parts. The opening frames the business risk or opportunity without any technology reference: 'NorthArc Manufacturing experienced eight unplanned production outages in 2024 with a combined impact of $2.7 million. Three were preceded by detectable signals our current monitoring infrastructure did not capture.' The problem statement is complete before the technology solution is introduced. This ordering matters: it establishes a problem worth solving before the committee is asked to evaluate the proposed solution.

The second part presents pilot evidence, not technology or architecture: 'The Dayton pilot monitored two compressor trains for six months. It produced two confirmed early detections, documented in the attached evidence book. Cost avoided: $612,000. Pilot cost: $310,000.' The third part presents the financial model: costs, benefits, payback, NPV, and scenarios. The fourth covers risk: what could prevent value delivery, likelihood, and mitigation. The fifth states the decision requested: funding amount, timeline, governance structure, and pre-defined Phase 2 gate criteria. Each part should be readable in under 4 minutes by a board member who has not been briefed on the program.

Diagram 4.8 - Five-Part Executive Investment Narrative Framework

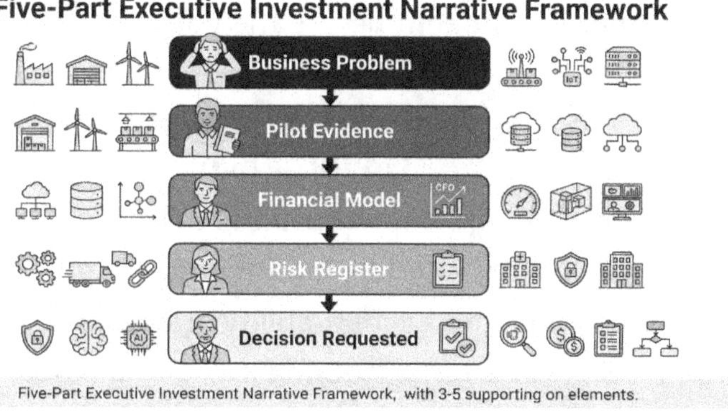

Five-Part Executive Investment Narrative Framework, with 3-5 supporting on elements.

Renata and Mei Lin rehearsed the NorthArc presentation with three internal reviewers: the General Counsel, who challenged every risk statement; the VP of Finance Planning, who stress-tested assumption disclosures; and Hector Salinas, who verified that operational claims matched the actual Dayton experience. Each produced a specific change. The General Counsel added a data-sovereignty risk that the program team had not fully considered. The VP of Finance Planning required a cleaner separation between firm costs and contingency. Hector corrected a maintenance cost figure that used an older estimate rather than the actual 2024 work order data.

Those changes made the business case more accurate and more defensible. They produced a secondary benefit: three senior leaders who helped refine the document had an ownership stake in its credibility. When the board asked whether the operational claims were sound, Hector Salinas answered from experience. That is the organizational mechanics of a fundable business case: the right people inside the tent before the vote.

Managing the Transition to Program Governance

Approval is not the end of the business case discipline. The assumptions, metrics, and commitments in the funded document become the accountability framework for the program. Every quarterly steering review should report actuals against the original projections. Variance requires explanation: not rationalization, but root-cause analysis and a revised estimate. Finance committees will accept variance; they will not accept unacknowledged and unmanaged variance.

Renata established a business-case tracking dashboard reporting four metrics at every steering committee meeting: cumulative program cost versus budget, cumulative value delivered versus projection, operator adoption rate, and model accuracy index. The first three tracked financial accountability. The fourth was an early warning indicator: declining model accuracy predicts declining value delivery within one to two quarters, allowing the governance team to intervene before the financial impact materializes.

Diagram 4.9 - Business Case Tracking Dashboard: Four Governance Metrics

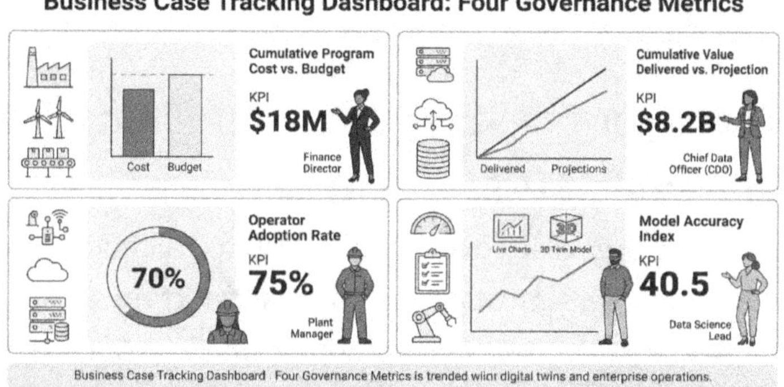

The transition from pilot to a funded program requires an upgrade in governance. An enterprise program spanning multiple operating units and committed to specific financial outcomes needs formal oversight: a steering committee with operating-unit representation, a defined escalation path for scope and budget decisions, a change-control process, and a regular performance-review cadence against business-case commitments. Document the governance design in the funding proposal and have it approved alongside the financial model, so that the committee authorizing the investment is also authorizing the oversight structure.

Programs funded without a governance structure tend to drift: scope expands, integration costs increase without benefit adjustments, and model quality degrades because no one owns the retraining workflow. In combination, these mechanisms produce the outcome that has damaged enterprise technology programs across industries: technically successful but financially disappointing, because the management infrastructure failed, not the technology.

5.8 Manager's Checklist: Building a Fundable Business Case

The following checklist covers the decisions, artifacts, and behaviors managers must own before submitting a digital twin business case for investment approval. A case in which every item cannot be checked is not ready for a finance committee.

- Baseline documented: five or more years of historical operational data (downtime frequency, maintenance cost, energy consumption, safety incidents) collected from your own systems and archived in auditable form.
- Value architecture map completed: each claimed benefit traced to a specific operational mechanism, a specific financial account, and categorized as hard savings, soft savings, or revenue effect.
- Double-counting reviewed: each dollar of claimed benefit verified to appear under exactly one value driver, with no overlap between avoided-downtime and energy-savings claims.

- TCO modeled across all five categories: platform licensing, integration and data infrastructure, model maintenance and governance, change management and training, security and compliance.
- Payback period calculated using the company's actual hurdle rate, with a benefit realization ramp that explicitly models deployment, adoption, and stabilization phases.
- Three-scenario NPV model completed: base case, conservative case (benefits minus 30 percent, costs plus 20 percent), and optimistic case.
- Sensitivity analysis identifies the two assumptions with the greatest financial impact on NPV; results shown as a two-variable grid.
- Pilot scoped to the minimum instrumented environment sufficient to observe and measure claimed value drivers within 6 to 12 months.
- Pre-defined success criteria negotiated with the approving executive before the pilot begins, with a Phase 2 funding gate tied to specific, measurable outcomes.
- Pilot instrumentation plan covers all metrics required by the business case: baseline metrics, twin performance metrics, and operational behavior metrics.
- Evidence book constructed: 10 or more documented pilot events, each with date, alert detail, operational response, confirmed outcome, and cost consequence.

- Funding pathway defined: CapEx versus OpEx split documented, accounting treatment for each element confirmed with finance, and shared-service allocation methodology designed if applicable.
- Integration contingency included: readiness assessment completed at each target facility, cost estimate reviewed by a technically qualified reviewer, and a 20 to 35 percent contingency reserve included as a named line item.
- Operating unit adoption commitments secured: each operating unit leader has signed a commitment naming the specific behaviors the program requires and the measurement approach.
- Executive narrative structured in five parts: business problem, pilot evidence, financial model, risk register, and decision requested.
- Pre-submission reviewers engaged: at least one finance reviewer, one operational reviewer, and one risk reviewer.
- Program governance structure included in the funding proposal: steering committee composition, escalation path, change control process, and performance review cadence.
- Business case tracking dashboard designed before approval: metrics, sources, reporting frequency, and owner defined for each tracked metric.

5.9 Takeaway

The business case for a digital twin program is a structured argument about operational cause and financial effect, built on documented baseline data, pilot evidence, and an honest accounting of the full cost of delivery. Programs funded on the strength of that argument tend to maintain the organizational support needed to reach the production scale at which real value materializes. Programs funded on vendor benchmarks and optimistic modeling tend to face a credibility crisis in year two, at exactly the moment when the hard integration work is done, and returns begin to arrive.

Renata Vance's board presentation was approved unanimously. Margaret Chen approved the funding with two conditions: the integration contingency reserve had to be maintained as a separate budget line, not absorbed into the base program budget, and the Phase 2 gate criteria had to be reported to the audit committee before Phase 2 spending began. Renata already planned both conditions. The CFO's conditions were the governance structure confirming itself.

For managers preparing their own cases: the goal is not to make the numbers look impressive. It is to make them look honest. Finance committees approve honest numbers with risk-mitigation plans. They reject optimistic numbers without acknowledging risk, even when the optimistic numbers are higher. The disciplines of assumption disclosure, scenario modeling, integration contingency, and pre-defined governance do not weaken a digital twin business case. They are what turn a promising pilot result into a funded program.

6 The Living Record: Data Architecture Decisions That Keep Twins Alive

6.1 Twin Sprawl: When Three Units Build Three Incompatible Worlds

Dr. Mei Lin Park had seen the slide before entering the conference room, forwarded by Renata Vance with a single line: 'We need to fix this before it becomes a board-level embarrassment.' The slide was a network map showing three separate data pipelines, three separate time-series stores, and three separate ontologies, each claiming to represent some part of NorthArc's operational reality. NorthArc Manufacturing had built its twin around the PI System historian. NorthArc Energy Services was streaming SCADA data through Apache Kafka into InfluxDB. NorthArc Logistics had engaged an integrator who delivered a medallion-architecture lakehouse on Azure with a proprietary asset schema that shared no vocabulary with either of the others.

Mei Lin called the situation 'twin sprawl': three data models each claiming to represent the same enterprise reality, unable to exchange a single attribute without a bespoke translation layer. When Renata asked a straightforward question, 'What is the current heat-stress index for every asset under active maintenance across all three divisions?' the answer required manual reconciliation across all three systems and took four days. That delay was the direct cost of using three silos instead of a single federated architecture.

What Mei Lin proposed was not a rip-and-replace consolidation. Hector Salinas would reject any plan that requires the Dayton plant to discard a PI System historian that has been accumulating calibrated sensor history for eleven years. Her proposal was a federated data fabric: a shared semantic layer, a shared identifier strategy, and governed data contracts, sitting above each division's existing store and translating between them through standard interfaces. The proposal took two months to scope and six months to implement in its first phase.

Diagram 5.1 - NorthArc Twin Sprawl Before and After Federated Data Fabric

NorthArc Twin Sprawl Before and After Federated Data Fabric

Before Federated Data Fabric

After Federated Data Fabric

Federated Data Fabric unifies siloed data for streamlined operations and clear insights.

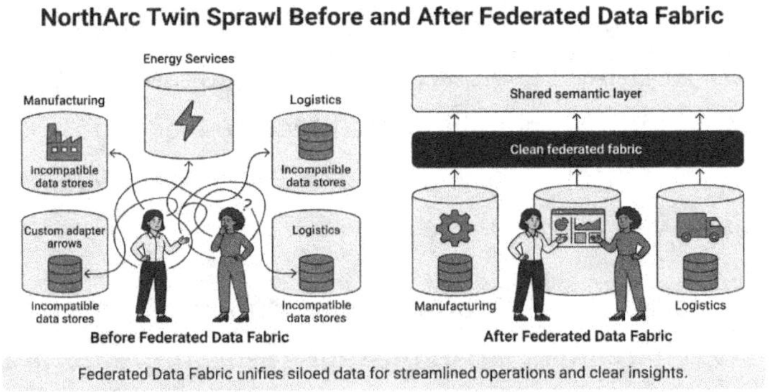

Every data architecture decision for a digital twin ultimately answers one question: how do we ensure the model reflects the physical world as it is right now, and

that we can trust what it says? That question breaks into five sub-problems addressed in this chapter. Identity: How does the twin know what it is modeling? Ingestion: how does real-time data flow reliably from sensors to the model? Storage: What persistence patterns serve the distinct read and write demands of twin workloads? Contracts: how do we govern the interfaces between source systems and the twin so that schema changes do not silently corrupt the model? Lineage: how do we know when the data is wrong, trace the source, and replay history to recover?

6.2 Asset Taxonomy and the Identifier Strategy That Holds Everything Together

Why Identity Is the Root Problem

The first question any twin must answer is: What am I modeling? The answer seems obvious until you discover that the same physical pump appears in your ERP as PMP-0447, in your CMMS as P-Dayton-CW-12, in your SCADA historian as TAG_COOLWATER_PUMP_12, and in your twin platform as a vendor-assigned GUID: four systems, four identifiers, one pump. Every pipeline joining information about that asset must resolve this mapping. Done ad hoc, the same reconciliation logic accumulates in a dozen places. When the CMMS is upgraded and the identifier changes, you find all twelve places at once, usually during an incident.

The solution is a canonical identifier that every system treats as the authoritative reference. Mint a persistent, globally unique identifier for each in-scope asset, register it in a lightweight asset registry, and require every downstream system to map its internal identifiers to the canonical ID. Priscilla Okonkwo discovered the lock-in dimension during procurement: the IoT integrator NorthArc was evaluating, proposing to assign identifiers from the vendor's own namespace. Mei Lin flagged this immediately. The contract was revised to require NorthArc to retain namespace ownership, with vendor-internal IDs treated as implementation details.

Building the Asset Taxonomy

An asset taxonomy defines the classification hierarchy that the twin uses to organize its knowledge. At NorthArc, Mei Lin's team settled on four levels: enterprise, facility, functional location, and asset instance. The taxonomy serves two purposes: attributes can be inherited downward through the hierarchy (a functional location carries nominal operating pressure and safety class that all its assets inherit), and the hierarchy scopes twin computations (a simulation of the cooling water loop needs to know which assets belong to it).

The Asset Administration Shell (AAS), standardized in IEC 63278, provides a widely adopted framework for asset identity and taxonomy, with versioned submodels covering nameplate data, technical properties, operational parameters, documentation references, and event history.

Whether your organization adopts AAS formally or builds an equivalent in a custom registry, the underlying principle is the same: an asset has a structured identity that specifies which properties it carries and who is responsible for maintaining them. When an asset is decommissioned, its canonical ID is retired rather than reassigned. When replaced, the successor receives a new ID, and the succession relationship is recorded, preserving historical data linkage while keeping the active model clean.

Diagram 5.2 - Four-Level Asset Taxonomy with Canonical Identifier Mapping

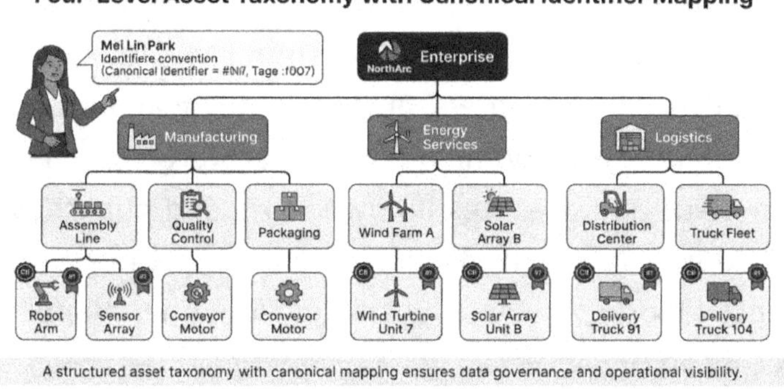

A structured asset taxonomy with canonical mapping ensures data governance and operational visibility.

6.3 Data Fabric vs. Data Mesh: Choosing the Integration Architecture

What a Data Fabric Does and When It Fits

A data fabric provides a unified access layer over distributed data stores without requiring physical consolidation. It knows where data lives, what it means, how fresh it is, and what access controls apply, and presents a consistent API surface to consumers. The fabric maintains a metadata catalog that maps each store's schema to a shared semantic vocabulary, provides governed-access APIs that enforce data contracts, and handles translation between local identifiers and canonical asset IDs.

A data mesh, by contrast, distributes ownership of data products to the domains that produce them. Each domain publishes its own product through standardized interfaces, and there is no central integration team. Mesh works well when domain teams are mature and have strong data engineering capabilities. It struggles when domain maturity varies or when centralized governance is needed to enforce quality standards. At NorthArc, Mei Lin chose a hybrid: a centrally governed semantic catalog and identity layer combined with domain-owned data products. Each division owns and operates its own store and publishes a data product conforming to the enterprise data contract. The central fabric provides the catalog, identifier mapping, and access governance.

Diagram 5.3 - Data Fabric vs. Data Mesh Architecture Comparison for Twin Programs

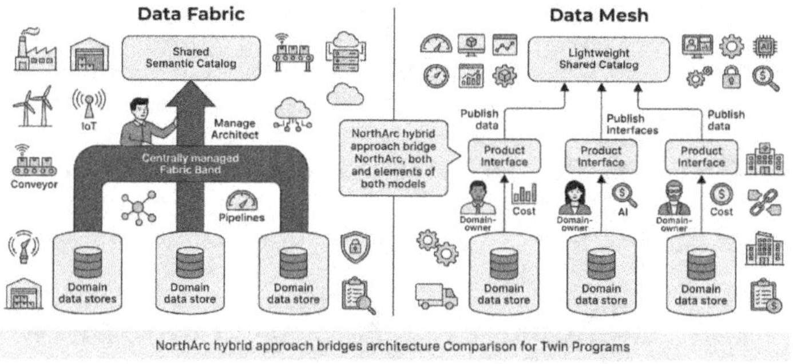

NorthArc hybrid approach bridges architecture Comparison for Twin Programs

Semantic Standards: ISO 23247 and the AAS

ISO 23247 defines a reference architecture for manufacturing digital twins. Its practical contribution to data architecture is the separation between the device connectivity layer (OPC UA adapters, MQTT brokers, edge gateways), the entity management layer (asset registry, twin state store), and the digital twin element layer (simulation models, analytics). When these three layers are conflated in a monolithic implementation, schema changes in the device layer propagate directly to the simulation layer, causing silent model corruption.

Managers should require that any proposed twin architecture can explicitly map its components to these three layers and articulate the contract between them. If a vendor cannot answer that question clearly, their architecture likely conflates concerns in ways that will produce data quality problems as the system matures. The standard provides the language for asking; the vendor's answer reveals the maturity of their design.

6.4 Storage Patterns: The Lakehouse, Time-Series Stores, and Where Twin Data Lives

The Lakehouse Pattern and Its Role in Twin Workloads

Digital twin workloads have an unusual storage profile: high-frequency, low-latency writes at 100-millisecond to 5-second intervals during ingestion, and analytical queries spanning months or years of history plus real-time API serving during consumption. No single storage system optimizes for all of these patterns. The lakehouse architecture addresses this through layering: an object store holding raw columnar data at the base, and an open-table format layer (Delta Lake or Apache Iceberg) above it, which adds ACID transactions, schema evolution support, time-travel queries, and incremental processing.

At NorthArc, the Energy Services division had already built a Delta Lake lakehouse for grid analytics. When Mei Lin's team evaluated it for the federated fabric, two challenges emerged: Delta's write latency was unsuitable for 100-millisecond vibration sensor streams, and the existing tables lacked indexed access patterns for real-time state lookups. The result was a two-tier design: a fast time-series store for high-frequency sensor data, and the Delta

Lakehouse for historical analytics and simulation input. The medallion architecture disciplines the raw data: bronze holds raw ingested data exactly as received, silver holds cleaned and validated data, and gold holds aggregations and feature sets ready for analytics and simulation.

Time-Series Stores and Retention Policy

Time-series databases (InfluxDB, TimescaleDB, OSIsoft PI System) are purpose-built for timestamped measurement workloads. The critical management decision is retention and resolution policy, not which store to choose. High-resolution sensor data sampled at 100 milliseconds accumulates roughly 35 gigabytes per sensor per year. A tiered strategy down samples to one-minute averages after 90 days and one-hour averages after one year, retaining full resolution only in cold archive for forensics, reducing ongoing storage costs by roughly 98 percent. Retention policy should be set at the asset class level, not as a uniform enterprise rule. Tomás Reyes objected to the standard 90-day window because the vibration anomaly detection model predicting bearing failures required full-resolution signatures going back 180 days. The resolution was targeted: vibration sensors on predictive maintenance assets retain full-resolution data for 365 days. Use case requirements, not storage economics alone, must drive retention policy.

Diagram 5.4 - Two-Tier Storage Architecture: Time-Series Fast Lane and Delta Lakehouse

Two-Tier Storage Architecture: Time-Series Fast Lane and Delta Lakehouse

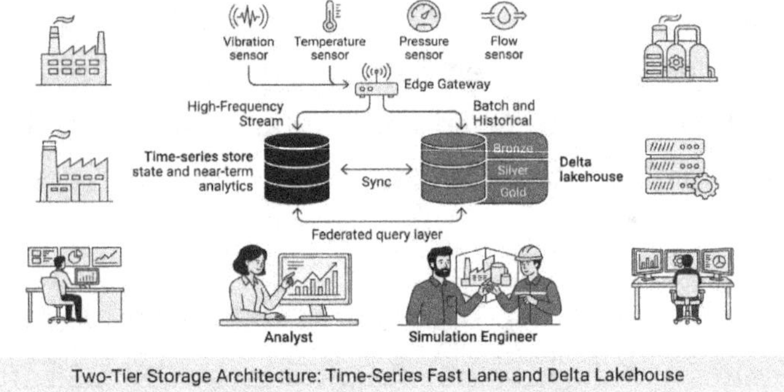

Two-Tier Storage Architecture: Time-Series Fast Lane and Delta Lakehouse

6.5 Data Contracts: The Agreement Between Source Systems and the Twin

What a Data Contract Is and Why It Matters

A data contract is a formal, versioned agreement that specifies what a data producer will deliver to a consumer: schema, latency guarantees, freshness expectations, quality commitments, and the process for notifying the consumer of breaking changes. Data contracts shift governance from reactive (the consumer discovers a schema change when the pipeline breaks) to proactive (the producer commits to a change-management process that gives the consumer time to adapt). For digital twin programs, where a silent schema change in an upstream SCADA system can corrupt a simulation model without triggering an error, this governance shift serves as a reliability mechanism.

At NorthArc, Mei Lin's team implemented contracts for the 27 source systems feeding the federated twin fabric. Each contract specified: the canonical asset IDs covered, the attributes delivered and their data types, the expected update frequency, the acceptable staleness tolerance, the payload format and encoding, and the change notification SLA (advance notice required before a breaking schema change). Contracts were stored in a version-controlled repository and referenced by each ingestion pipeline as the authoritative schema definition. Renata Vance made data contract compliance a quarterly metric on the steering committee dashboard: the percentage of source systems with active contracts and the percentage meeting their change notification SLA. Tying compliance to executive visibility proved more effective than any technical enforcement mechanism.

Schema Evolution and Registry Governance

Without governed schema evolution, a producer who renames a field breaks every consumer referencing the old name. A type change from integer to float causes silent data corruption in any consumer that casts the value without type checking. The standard approach relies on a schema registry: a centralized store of schema definitions enforcing compatibility rules. Apache Schema Registry (commonly deployed alongside Kafka) supports backward compatibility (new schemas can read data written with old schemas), forward compatibility (old schemas can read data written with new schemas), and full compatibility (in both directions). Backward compatibility is the minimum requirement for twin ingestion pipelines.

Aisha Bramwell identified data contracts as a risk management instrument. A twin used for regulated reporting must demonstrate that its input data was received with known provenance and a validated schema. A contract audit trail showing which schema version was active during which time window provides exactly the traceability a regulatory examiner needs. Aisha's team added a contract audit log requirement to NorthArc's data governance policy: schema version history retained for at least seven years for any twin feeding regulated reporting.

Diagram 5.5 - Data Contract Lifecycle from Source System to Twin Ingestion Pipeline

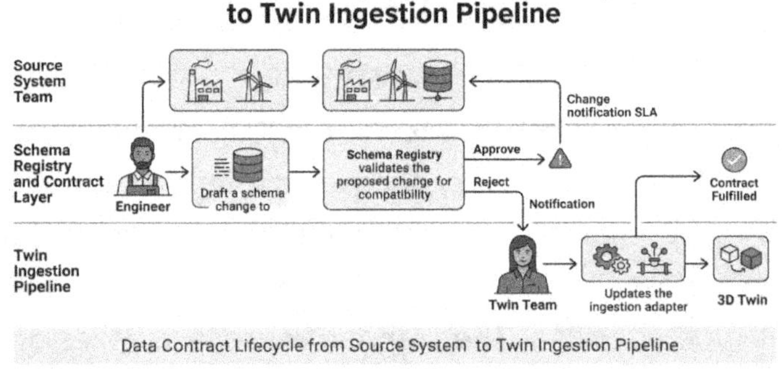

6.6 Master Data Management: Keeping the Twins' Reference Data Clean

Four Categories of Reference Data

Master data management for digital twins covers four categories: asset master (canonical registry of physical assets, identifiers, classification, and lifecycle status), location master (physical and logical locations structured to match the asset taxonomy), maintenance master (maintenance procedures, spare part specifications, and manufacturer documentation per asset class), and personnel master (roles and qualifications relevant to operating and maintaining each asset). Without clean reference data in these four categories, a twin with excellent sensor feeds still lacks contextual grounding. It can report that asset PMP-0447 exceeded its vibration threshold, but cannot tell you what type of pump it is, where it sits in the process, what the maintenance procedure requires, or who the qualified technician on duty is.

The asset master is the most critical and most commonly neglected. Enterprises that have grown through acquisition carry multiple CMMS systems and classification schemes that do not align. NorthArc's acquisition of its energy services division added a second CMMS (IBM Maximo alongside SAP Plant Maintenance) and a second asset classification standard. Reconciling these into a unified asset master took six months and consumed more effort than any single technical component of the twin implementation. Managers who have not completed this reconciliation before launching a twin program will do it during the program, under pressure, with live integrations already depending on unreconciled data.

The Semantic Vocabulary Problem

Beyond structure, a twin needs a semantic vocabulary: a defined set of terms with agreed meanings. What does 'operational' mean for an asset? Powered on? Actively processing? Available to process if needed? These distinctions matter because a predictive maintenance model trained on data labeled 'operational' will behave incorrectly if the label is applied inconsistently across divisions.

At NorthArc, Mei Lin's team used ISO 23247 as the foundation for the manufacturing twin vocabulary, the IEC Common Information Model for energy services, and a custom extension layer for logistics, all mapped to a shared enterprise ontology maintained in Git with a review process for additions. The practical test of semantic alignment is whether a cross-domain query returns meaningful results without manual post-processing. When Renata Vance's heat-stress index query had required four days to return on the fabric, it returned in under 30 seconds afterward. The improvement was not in computational speed. It was the elimination of manual translation work.

Diagram 5.6 - Master Data Management Layers Supporting Digital Twin Reference Data

Master Data Management Layers Supporting Digital Twin Reference Data

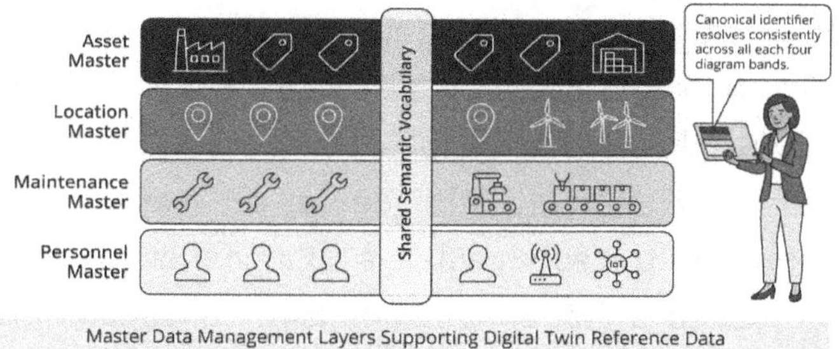

Master Data Management Layers Supporting Digital Twin Reference Data

6.7 Lineage, Observability, and the Quality Gates That Protect the Model

Data Lineage: Tracing Every Value to Its Source

Data lineage is the documented record of where a value in the twin originated: which physical sensor produced it, which edge gateway collected it, which ingestion pipeline transformed it, and which quality gate validated it. Lineage answers 'why does the model say this?' with a traceable chain back to a physical measurement. Without lineage, an unexpected simulation result requires investigation from scratch. With lineage, the investigation starts from the last confirmed good value and traces forward through the transformation chain.

Apache Atlas, OpenLineage, Delta Lake's audit log, and Iceberg's metadata layer all provide mechanisms for capturing lineage. The key is treating lineage capture as a required element of pipeline design, not an afterthought.

Every transformation step from the sensor to the twin model should emit a lineage event that includes the input dataset version, the output dataset version, the applied transformation, the quality-gate result, and a timestamp. At NorthArc, the lineage graph proved its value during a grid incident: an anomaly in the energy services twin's load forecasting was traced via lineage to a new edge gateway that introduced a Fahrenheit-to-Celsius conversion error. Because the error changed no schema field, schema validation had not caught it. Lineage tracing identified the root cause in minutes rather than hours.

Observability and Quality Gates

Observability for a digital twin means treating the data pipeline as an operational system with its own health metrics and alerts. The four key metrics are freshness (age of the most recent value relative to the expected update frequency), completeness (percentage of expected sensor readings that arrived in the last interval), accuracy (percentage of incoming values that pass quality-gate validation), and latency (time from the physical event to the twin-model update). Each should have a defined threshold and an alert when exceeded. A sensor that has not sent a reading in 10 minutes when its expected frequency is 5 seconds is either physically or connectivity-failed. The twin should not silently treat the last known value as current. It should mark the sensor point as stale and propagate that uncertainty to any model output depending on it.

Dorian Whitlow insisted on a data quality dashboard visible alongside the twin visualization in Dayton's control room. Tomás Reyes initially objected. After a three-month pilot, he reversed his position: operators learned to cross-check freshness indicators before escalating anomalous readings, reducing nuisance maintenance calls at the plant by approximately 18 percent. Quality gates enforce validation rules before committing data to the silver or gold tier: range checks, rate-of-change checks, reference validity checks, and timestamp coherence checks. Records that fail a quality gate go to a dead-letter quarantine for manual review rather than being silently dropped or passed. The quarantine creates a visible backlog that drives the feedback loop needed to improve sensors and validation rules alike.

Diagram 5.7 - Data Observability Dashboard: Freshness, Completeness, Accuracy, and Latency Metrics

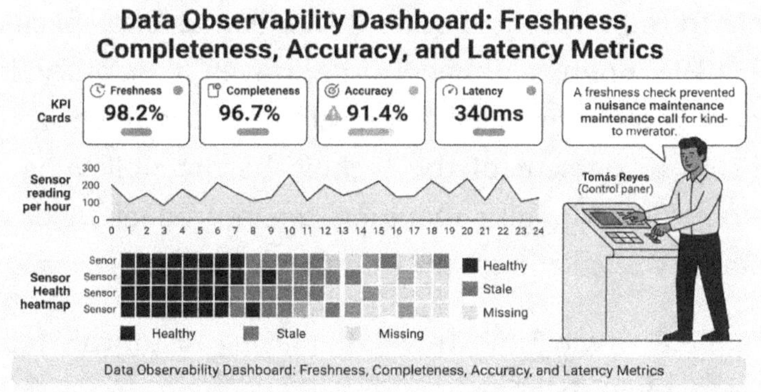

Data Observability Dashboard: Freshness, Completeness, Accuracy, and Latency Metrics

6.8 Retention, Replay, and Twin Time Travel

Three Use Cases for Time Travel

Time travel in digital twin data management is the ability to reconstruct the full state of the twin at any point in the past: not just querying a historical sensor reading, but reconstructing asset states, relationship states, simulation parameters, and data quality context as they existed at a specific timestamp. Three use cases drive this requirement. Incident investigation: replaying the twins' state in the hours leading up to a failure lets engineers identify leading indicators, test alternative intervention scenarios, and produce a documented timeline for post-incident review. Model validation: new simulation models must be backtested against historical periods where the outcome is known before being trusted in live operations. Regulatory audit: a twin feeding regulated reporting must be able to reproduce a specific calculation as of a specific date. Aisha Bramwell made time travel mandatory in NorthArc's platform evaluation criteria, assigning it a weight of 15 percent of the technical score, following a prior regulatory inquiry requiring the reproduction of a grid stability calculation from 14 months earlier.

Diagram 5.8 - Twin Time Travel: Replaying Historical State for Incident Investigation and Model Validation

Twin Time Travel: Replaying Historical State for Incident Investigation and Model Validation

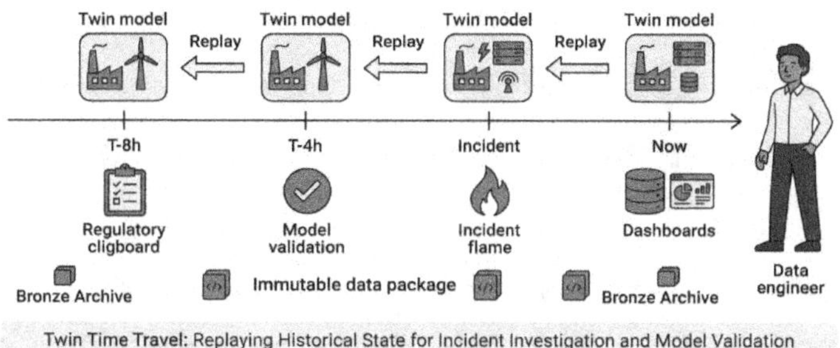

Twin Time Travel: Replaying Historical State for Incident Investigation and Model Validation

Implementing Time Travel: Delta, Iceberg, and Retention Policy

Delta Lake and Apache Iceberg both provide native time travel through transaction log and snapshot metadata. The time-travel window is limited by the length of time the transaction log is retained. For the regulatory audit use case with a seven-year retention obligation, Mei Lin's team implemented a hybrid: the active transaction log is retained for 90 days, with quarterly full snapshots of the gold tier retained for seven years in cold storage.

Managers must verify that the time travel mechanism has been tested, not just documented. A retention policy that exists on paper but has never been exercised is a theoretical capability that will fail under incident pressure. NorthArc's platform team ran a quarterly time travel drill: pick a random historical date, reconstruct the twin state for that date, and verify the result against the bronze-tier raw archive. The first drill took three days. By the fourth, it took two hours. Repeated testing converts documentation into operational capability.

6.9 What Managers Must Demand: Preventing Twin Sprawl Before It Starts

The NorthArc scenario was recoverable. Priscilla Okonkwo estimated the federated fabric remediation at approximately $2.1 million in consulting and internal effort over eight months. Each division's implementation, while incompatible with the others, was internally consistent. What was missing was the cross-division governance layer. Organizations that discover twin sprawl after each division has selected a different vendor platform with incompatible data models and proprietary identifier schemes face substantially higher remediation costs. The options narrow to either maintaining permanent translation layers (indefinitely accumulating integration debt) or consolidating onto a single platform (requiring renegotiation of vendor contracts and reimplementation of working integrations). The governance investment that prevents this situation costs an order of magnitude less.

The governance requirements that prevent twin sprawl are organizationally demanding rather than technically complex. They require that decisions that business units currently make autonomously, such as which historian to use, how to name assets, and what schema to publish, become shared decisions governed by a cross-functional architecture body. Renata Vance chartered the NorthArc Digital Twin Architecture Board, chaired by Mei Lin, with representatives from each operating division plus IT, legal, and risk. The board met monthly, approved data architecture decisions for new deployments, and reviewed existing deployments annually. It had explicit authority to delay a deployment that did not meet the federated fabric standards, with escalation rights to Renata.

The five governance decisions the board controlled were: identifier namespace, semantic vocabulary, storage tier selection, data contract format, and change notification SLA. Consistently enforced, these five decisions prevented additional incompatible silos while preserving enough local autonomy that division teams did not feel micromanaged. Architecture Decision Records (one page per decision in a Git repository) provided institutional memory for new team members and external auditors. Priscilla Okonkwo added data portability requirements to all vendor contracts: documented export API in an open format, identifiers in NorthArc's namespace, and commitment to the standard change notification SLA. Three of seven vendors in NorthArc's evaluation failed the identifier namespace requirement.

Diagram 5.9 - Digital Twin Architecture Governance Board: Decision Authority and Escalation Paths

Digital Twin Architecture Governance Board: Decision Authority and Escalation Paths

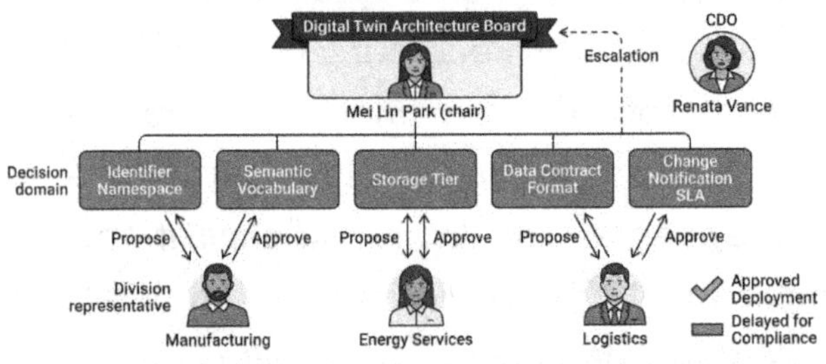

STRUCTURED GOVERNANCE FOR EFFICIENT DIGITAL TWIN ARCHITECTURE DECISIONS.

6.10Manager's Checklist: Data Architecture Decisions to Own Before Deployment

The items below capture the decisions, artifacts, and behaviors a manager overseeing a digital twin program must own or actively sponsor.

Identity and Taxonomy

- Establish a canonical identifier namespace owned by your organization, not any vendor. Verify new deployments use this namespace before contracts are signed.
- Define and publish a four-level asset taxonomy before any twin deployment begins. Deviations require Architecture Board approval.
- Confirm your CMMS and ERP asset records can be reconciled to the canonical taxonomy. Budget MDM remediation as a prerequisite, not a follow-on activity.
- Assign explicit ownership of the asset master to a named Data Owner with authority to adjudicate identifier conflicts and approve taxonomy changes.

Data Contracts and Schema Governance

- Require signed data contracts for every source system feeding the twin, covering schema version, update frequency, freshness SLA, and change notification obligations.
- Deploy a schema registry with at least backward compatibility enforcement before any production ingestion pipeline goes live.
- Include data contract compliance as a standing metric in program steering reviews: percentage of source systems with active contracts and percentage meeting change notification SLA.
- Require contract audit log retention for at least as long as your longest regulatory obligation.

Storage and Retention

- Select storage tiers deliberately: high-frequency sensor data to a time-series store, historical analytics and simulation inputs to a Delta or Iceberg lakehouse with medallion architecture. Document the rationale in an Architecture Decision Record.
- Define retention and downsampling policies per asset class based on use case requirements, not as uniform enterprise rules.
- Test time travel capability in a scheduled quarterly drill. A replay that has never been exercised is a theoretical capability, not an operational one.

Lineage, Observability, and Quality

- Require lineage capture at every pipeline transformation step as a pipeline design standard, not an optional feature to be added post-launch.
- Define the four key observability metrics (freshness, completeness, accuracy, and latency) with explicit alert thresholds. Make these metrics visible in the operational environment, not only in a back-office dashboard.
- Establish a dead-letter quarantine for records failing quality gates, with a weekly review process and feedback loop to source system teams. Silent dropping of failed records is not acceptable.
- Run a data quality review as a standing agenda item in program governance, not as a reactive response to incidents.

Integration Architecture and Sprawl Prevention

- Charter a cross-functional Architecture Board with authority to approve new twin deployments and review existing ones annually. Chair at the director or VP level with escalation rights to the program sponsor.
- Limit the approved technology list to what the enterprise can support reliably. Novel choices require explicit board approval with a documented supportability rationale.
- Include data portability requirements in all twin platform vendor contracts: open export API,

organization-namespace identifiers, and standard change notification SLA. Inability to meet these requirements disqualifies procurement.

- Conduct an annual twin data architecture audit reviewing each deployment against the approved reference architecture. Publish results to program leadership with a remediation timeline for any deviations.

6.11Takeaway

The living record that a digital twin depends on does not maintain itself. Every value in the model traces back to a physical sensor, travels through an ingestion pipeline, passes a quality gate, lands in a storage tier, and arrives at the model with a latency, freshness, and accuracy profile that determines whether the model reflects reality or a degraded approximation of it. The decisions governing this chain, identity, contracts, storage, lineage, and governance structure, carry organizational consequences and must be owned by the manager accountable for the twins' operational accuracy.

Twin sprawl is the most common failure mode for enterprise twin programs in their second and third year. It begins with reasonable local autonomy and ends with expensive, incompatible silos. The antidote is a federated architecture governed by a small set of non-negotiable standards: a shared identifier namespace, a shared semantic vocabulary, governed data contracts, and a cross-functional body with authority to enforce these standards before a new deployment goes live. The governance structure must be established before the second twin deployment, not after the third.

The manager who invests in these foundational decisions at the beginning of a twin program will spend the next three years building on a stable data foundation. The manager who defers them in favor of faster initial deployment will spend those years remediating architectural debt. The simulations, predictions, and operational insights that create value from a digital twin will not work reliably without the data architecture underneath them working well. Build the foundation first, and own it at the management level where the organizational authority to enforce cross-division standards actually resides.

7 The Thinking Twin: AI, Predictive Simulation, and Agentic Workflows

7.1 Scenario: The Prescriptive Overlay

The quarterly steering committee at NorthArc Industries convened in Columbus on a Thursday morning in late March. Renata Vance had blocked two full hours because Dorian Whitlow had requested time to demonstrate a prescriptive AI overlay on the Dayton plant twin. Hector Salinas sat near the front, visibly skeptical. Aisha Bramwell positioned herself at the corner with a notepad. The screen displayed the familiar three-dimensional model of Dayton's fabrication line, sensor feeds scrolling in the sidebar.

Dorian opened with a brief framing. The Dayton twin had been running in predictive mode for eleven months, flagging equipment likely to degrade within a thirty-day window. What his team had built over the past quarter was the next layer: a model that not only says something may fail but also tells the operator what to do, in what sequence, given current production targets, crew availability, and parts inventory. The interface now showed a ranked list of interventions with expected outcome, confidence level, estimated downtime cost, and a one-click path to scheduling the work order.

Hector asked what happened when two recommendations conflicted, and whether the model knew about Dayton's union work-rule window. Aisha asked how the audit trail worked if a supervisor acted on a bad recommendation. Renata watched both exchanges, then asked the question that set the tone for the rest of the meeting: at what point does the system stop recommending and start acting on its own, and who approved that transition? That question sits at the center of this chapter.

Diagram 6.1 - The Four Behavioral Modes of a Digital Twin

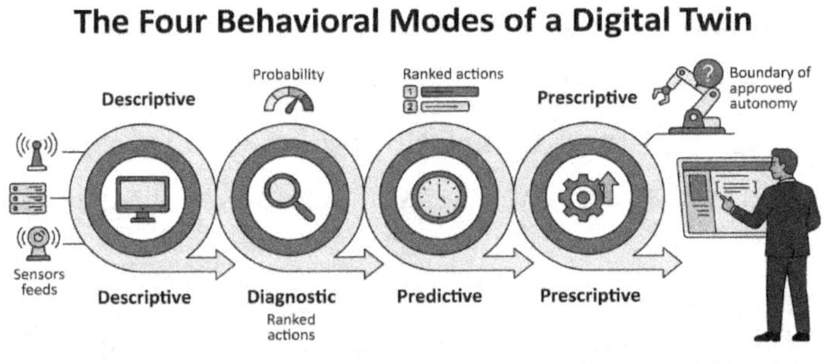

The four behavioral modes of a digital twin is : the boundary of approved autonomy.

The progression from descriptive to autonomous is a governance transition as much as a technology upgrade. Each step shifts more decision weight from a human to a model, requiring corresponding obligations to validate the model, define authority limits, and establish conditions for mandatory human override. Managers who understand this progression set appropriate boundaries—those who do not tend to either over-restrict or under-govern their AI investments.

This chapter maps that progression in practical terms: modeling techniques that underpin prediction and simulation, surrogate models and physics-informed neural networks, explained for non-engineers; anomaly detection and reinforcement learning as operational tools; generative scenario simulation; LLM copilots over twin context; and agentic workflows. Governance runs through every section, because the accountability framework must be built before autonomy is extended, not after.

7.2 Descriptive to Prescriptive: The Twin Behavior Spectrum

What Each Maturity Stage Actually Does

The twin maturity ladder runs through four named stages: descriptive, diagnostic, predictive, and prescriptive, with autonomous sometimes added as a fifth. A descriptive twin mirrors the current state based on sensor data. A diagnostic twin explains why something happened, correlating signals with known failure signatures. Value at these first two stages is real but bounded; the twin is a better instrument panel, not a thinking partner.

A predictive twin introduces time as a dimension. Rather than showing the current bearing temperature, it estimates the probability distribution of that temperature over the next 14 days, given operating conditions and historical patterns. Machine learning enters the architecture meaningfully here: the model learns which signal combinations tend to precede failures. NorthArc's

Dayton pilot ran in this mode for eleven months before earning enough operational trust to justify the prescriptive investment.

A prescriptive twin generates ranked action recommendations, not just probability scores. It encodes constraints (production schedules, crew windows, parts availability, regulatory intervals, risk tolerance) and produces the action sequence that best satisfies those constraints. Constraint encoding is itself a governance task. An autonomous twin goes further: it executes actions without per-step human approval. The governance requirements at the autonomous stage differ not in degree but in kind from earlier stages.

Why the Spectrum Matters for Budget and Timeline Decisions

Vendor proposals that promise to move a twin from descriptive to prescriptive in a single program increment are a warning sign. Each stage requires data quality, model validation, and trust-building that cannot be arbitrarily compressed. A prescriptive layer built on an undertested predictive layer inherits those errors as bad recommendations in high-stakes moments. NorthArc's program took 14 months to reach a reliably predictive state for the Dayton fabrication line, with a dedicated team and a focused scope. The prescriptive layer Dorian demonstrated had been in shadow mode for four months before any production use. Shadow mode is the period when the model earns operational trust and the governance team builds the oversight mechanisms that production deployment depends on.

Diagram 6.2 - Twin Maturity Stage: Investment and Governance Requirements

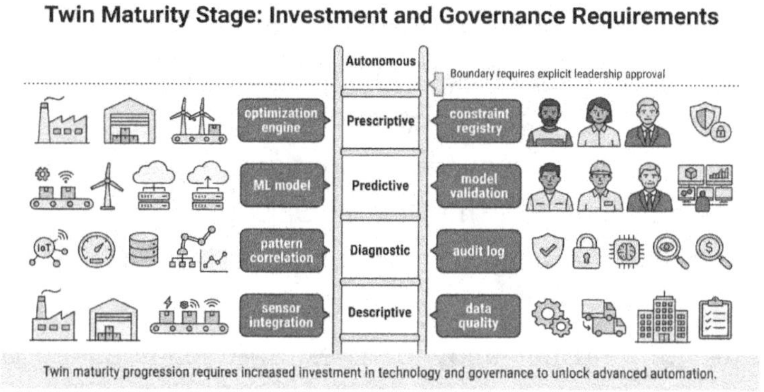

7.3 Surrogate Models and Physics-Informed Neural Networks: What Managers Need to Know

The Surrogate Model: Speed Without Sacrificing Fidelity

A full physics simulation of a manufacturing line or power grid is accurate but computationally expensive. Running it in real time, or across thousands of scenario variations, is prohibitively slow even on well-provisioned infrastructure. A surrogate model solves this: a lightweight machine-learning model trained to approximate the outputs of the full simulation, producing nearly equivalent results in orders of magnitude less time. Think of it as a skilled estimator who has studied the engineering blueprints thoroughly enough to give reliable answers in seconds rather than hours.

Surrogate models unlock real-time use cases that physics simulation alone cannot support. They introduce a distinct model risk: accuracy degrades when conditions move outside the training range. A well-governed surrogate documents its valid operating envelope and monitors for conditions approaching that boundary. Dorian's team trained a surrogate on the Dayton OEM physics model in three weeks, reaching recommendations in under two seconds. Tomás Reyes flagged that confidence should degrade for unusual alloy batches; the team added a conditional warning.

Physics-Informed Neural Networks: When Domain Knowledge Improves the Model

Standard machine-learning models learn purely from data. Physics-informed neural networks (PINNs) take a different approach: they encode known physical laws into the training process, so predictions must satisfy both observed data and the governing equations of the system, such as the thermodynamic constraints on turbine behavior. This constraint improves accuracy in data-sparse regions and makes the model's behavior interpretable, because predictions must be consistent with physics that a domain expert can reason about.

PINNs matter for managers for two reasons. First, they reduce training data requirements: a turbine that fails twice a year gives a purely data-driven model very little to learn from, whereas a PINN uses those two failures, along with physical equations, to produce a reliable model. Second, PINNs flag risk by pointing to physical mechanisms rather than statistical correlations alone, which matters when a recommendation influences a safety-critical decision. Many production architectures use both: a PINN as the high-fidelity core and a surrogate trained on PINN outputs for real-time decision support. Managers should ask which approach a vendor uses for each function and whether the team has relevant domain physics expertise.

Diagram 6.3 - Surrogate Models and PINNs in the Twin Architecture

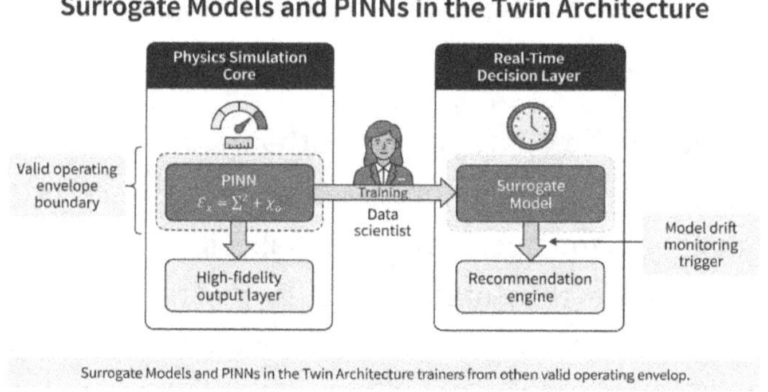

Surrogate Models and PINNs in the Twin Architecture trainers from othen valid operating envelop.

7.4 Anomaly Detection: The Twin as Vigilant Observer

How Anomaly Detection Works in a Twin Context

Anomaly detection is among the highest-value AI applications in a digital twin. The twin continuously compares the current sensor-derived asset state against the modeled baseline for similar conditions; deviations beyond a threshold trigger an alert. Most implementations combine rule-based thresholds for well-understood failure modes with statistical models for subtler patterns. Common methods include isolation forests, autoencoders, and multivariate control charts. Before configuring the Dayton anomaly layer, Dorian's team ran a joint workshop with Hector's maintenance engineers to map the costliest failure modes to the detection methods most sensitive to their early signatures.

Alarm fatigue is the operational risk that undermines these programs. When a twin generates dozens of alerts per shift, most benign, operators rationally ignore the alert pane. The metric that matters is the positive predictive value of the detection system: actionable alerts as a share of total alerts. A well-tuned layer for a critical asset should produce fewer than 5 alerts per week with a positive predictive value above 70%. Reaching that threshold requires months of operator feedback documenting whether each alert corresponded to a real degradation event.

Governing Anomaly Detection: Thresholds as Policy Decisions

Alert thresholds encode a risk policy, not just a technical setting. How sensitive should monitoring be, and what is the acceptable cost of false positives (unnecessary maintenance) versus false negatives (missed failures)? Managers must own this decision; it cannot be fully delegated to a data science team that lacks context on downtime cost, regulatory exposure, and risk appetite. At NorthArc, Aisha required that threshold configurations for safety-critical equipment be documented in a formal decision register, including the date, business rationale, and the approving manager's name. Changes require sign-off from both the plant manager and the risk team, preventing a well-intentioned engineer from quietly improving precision metrics at the cost of the sensitivity that the safety case depends on.

Diagram 6.4 - Anomaly Detection Governance: From Sensor to Decision

Anomaly Detection Governance: From Sensor to Decision

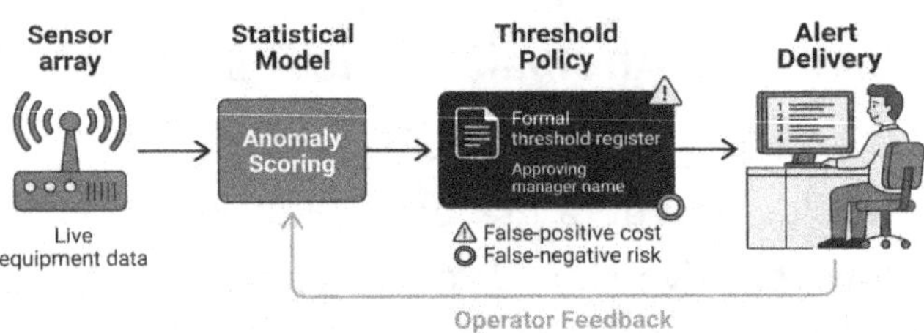

Anomaly detection governance : From **sensor to decision**, review ss, ateriliant on ranked alert list.

7.5 Generative Scenario Simulation: What Would Happen If...

Turning the Twin into a War-Gaming Environment

Generative scenario simulation lets a manager ask the twin a what-if question and receive a quantified answer within operational time frames. What happens to throughput if the third press line goes down for 72 hours during peak-demand week? What is the probability of distribution of on-time delivery if the primary supplier delays by ten days? Before AI-augmented twins, answering those questions required expert intuition or multi-day modeling exercises. A twin with a well-trained simulation layer answers them in minutes.

The technical underpinning is the ability to perturb initial conditions or constraint parameters and propagate perturbations through the model to generate output distributions. Monte Carlo methods run hundreds of simulated scenarios with randomized inputs drawn from calibrated probability distributions. Generative methods using neural networks extend this by surfacing scenarios the planner had not thought to test. The governance question is what decisions the simulation capability is meant to support. A well-governed tool is scoped to specific decision types, such as capital allocation, contingency planning, shift scheduling, and supplier risk assessment. An unscoped simulation tool generates interesting analyses that never connect to decisions, consuming data science resources without workflow-level impact.

LLM Copilots Over Twin Context: Operator Chat at the Twin Interface

LLM copilots over digital twin context are a recent and rapidly maturing capability. The twin maintains a continuously updated knowledge context: current asset states, recent alerts, maintenance records, operational parameters, and procedural documentation. An LLM interface lets an operator query that context in natural language rather than navigating three separate systems to find a compressor's last maintenance date, its current vibration reading, and the relevant inspection procedure. NorthArc piloted an LLM copilot on the Dayton twin after integrating the prescriptive overlay; Tomás Reyes's floor team found it most valuable during handoff shifts, when incoming operators needed to orient to the state of multiple assets quickly.

Three governance risks require attention. Hallucination must be bounded through retrieval-grounded design: the LLM synthesizes answers from retrieved twin data, not general training. Scope creep, where operators query outside the twin's validated data domain, must be managed through clear interface design and logging of out-of-scope queries. Accountability requires that every response be logged with the source data, timestamp, and user so that consequential actions can be traced to specific information the copilot provided.

Diagram 6.5 - LLM Copilot Architecture Over Twin Context

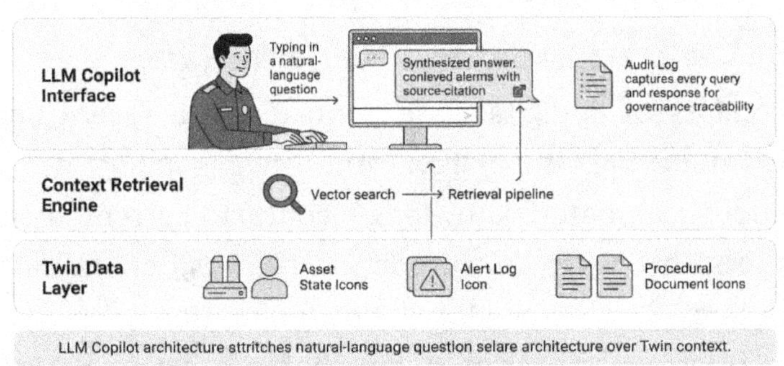

7.6 Reinforcement Learning for Control Optimization

What Reinforcement Learning Does in an Operational Twin

Reinforcement learning (RL) trains a control policy by having an agent take actions in an environment, receive reward signals, and iteratively improve the policy to maximize cumulative reward. In a digital twin context, the twin is the training environment. Rather than testing policies on the physical asset, where a bad decision could damage equipment, the agent trains against simulation, experiencing thousands of operational scenarios in the time it would take to observe a handful of real ones. Once a policy demonstrates robust simulated performance, it is deployed on the physical asset with designed guardrails.

Practical RL applications include energy optimization for HVAC and large facility systems, process parameter tuning in continuous manufacturing, and set-point management for utility distribution. These use cases share the traits that make RL tractable: a well-defined reward signal, a continuous action space, and a simulation accurate enough for a policy to transfer reliably. When those conditions are not met, RL produces policies that perform well in simulation and fail in production. NorthArc Energy Services learned this firsthand when its first grid-twin RL policy exhibited unexpected switching behavior during stress events. The simulation had not accurately represented legacy relay timing characteristics; closing that gap took 4 months, after which the policy could be redeployed.

The Simulation-to-Reality Gap: Managing Transfer Risk

A structured transfer protocol manages the gap: validate the simulation against historical operational data before

RL training begins, deploy conservatively with narrow authority limits, observe in shadow mode before full activation, and monitor explicitly for behavior outside the trained distribution. Managers should treat the simulation-to-reality gap as a model risk issue. Aisha's team required that any RL-derived control policy document three things before production deployment: the operating conditions under which the policy was validated, the conditions requiring mandatory human override, and the fallback control strategy if the policy is suspended.

Diagram 6.6 - Reinforcement Learning Training Loop Inside the Twin

Reinforcement Learning Training Loop Inside the Twin

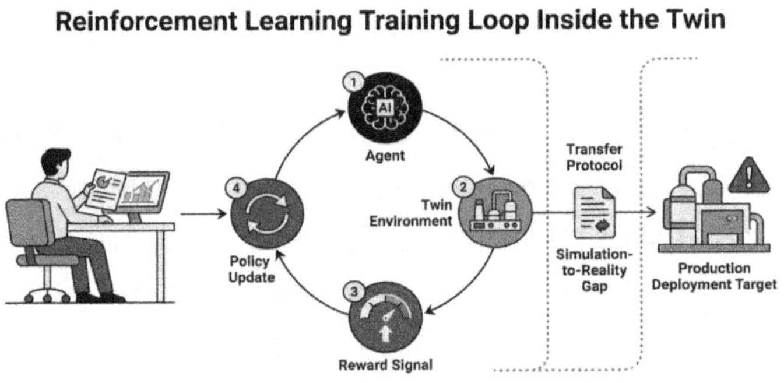

This infographic details the Reinforcement Learning process iterating within a digital twin before deployment to the physical asset.

7.7 Agentic Workflows: When AI Can Act, Not Just Recommend

The Architecture of an Agentic Twin Workflow

An agentic workflow in a digital twin context is a system in which an AI component can execute actions in the operational environment without requiring per-step human approval. The agent observes the twin's state, selects an action from its defined authority set, executes it through a system integration, and observes the outcome to inform the next cycle. Current enterprise implementations are almost always narrowly scoped: adjusting temperature set points within a defined range, triggering low-value purchase orders, or rerouting shipments between pre-approved carrier options. The key design principle is that the authority set must be explicitly enumerated. An agent authorized to take any action that optimizes energy costs has a boundary that is too vague to govern. An agent authorized to adjust the chiller set point between 44 and 48 degrees Fahrenheit during off-peak hours has a boundary specific enough to monitor and audit.

At the March steering committee, Dorian presented a scoped agentic proposal: an agent with authority to trigger standard preventive maintenance work orders for three specific asset classes at Dayton when model confidence exceeded 90%, and the work order value was below

$5,000. Renata asked about the override mechanism. Hector asked what happened when the agent triggered a work order conflicting with a planned production run. Aisha asked for the audit log specification before providing governance sign-off. None of the three questions had a complete answer yet. The committee approved a ninety-day shadow-mode trial.

Human-in-the-Loop Boundaries: What Must Stay with a Person

A useful framework organizes decisions along two dimensions: consequence severity (what is the worst plausible outcome if this decision is wrong?) and reversibility (how quickly and completely can the decision be undone if it is wrong?). Low-severity, highly reversible decisions are strong candidates for agentic authority. High-severity or irreversible decisions must remain with a human regardless of model confidence. Certain categories should never be delegated without a human checkpoint: regulatory notifications or compliance filings, safety system configuration changes, personnel decisions, capital commitments above defined thresholds, and any action affecting external stakeholders.

The graduated authority model is a practical structure: an autonomous tier where the agent acts without approval, a supervised tier where it acts but logs and reports within a defined window, and an approval tier where it routes through a human queue before acting. This avoids the binary choice between full autonomy and no autonomy and allows the organization to extend agentic capability incrementally as trust is built through operating history.

Diagram 6.7 - Agentic Authority Tiers and Human-in-the-Loop Boundaries

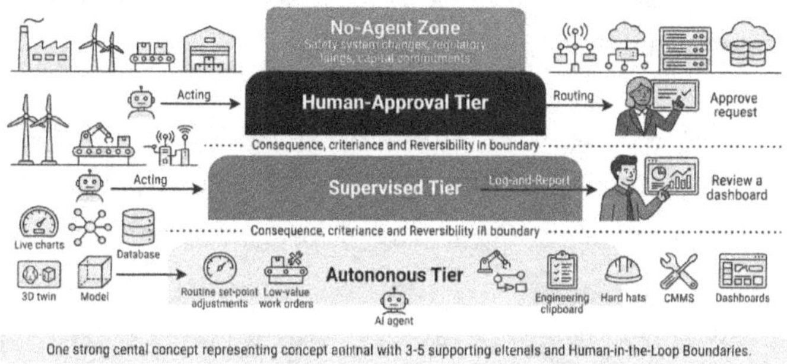

7.8 Model Risk Management for Twin AI

Applying SR 11-7 Style Controls to Operational AI Models

Model risk management (MRM) originated in financial services, where regulators codified requirements for validating, monitoring, and governing quantitative models. The Federal Reserve's SR 11-7 guidance established principles that have now been broadly adopted as AI model governance became a board-level concern. The core principles apply directly to digital twin AI: develop models with a clear use-case statement, validate them independently of the team that built them, deploy with documented operational limits, monitor for performance degradation, and retire through a controlled process. The model inventory is the foundation: every production AI model is listed with its use case, validation status, performance metric, model owner, and escalation path. Dorian maintains NorthArc's inventory on a platform accessible to Aisha's governance team. At the March steering committee, he reported that four of seven production models were due for six-month revalidation in the following quarter.

Model drift is the specific failure mode MRM monitoring is designed to catch. Equipment ages differently from the training data captured. Operational practices shift. Supply-chain disruptions alter raw material properties, affecting process behavior. Regular revalidation against fresh data, including adversarial testing designed to surface failure modes that routine validation would miss, is the mechanism that keeps drift from becoming an operational liability.

Explainability Standards for Non-Engineering Audiences

When an AI model's output influences a consequential decision, the responsible manager needs to understand the basis for the recommendation well enough to exercise meaningful oversight. That is not a demand for mathematical transparency; it is a demand for explanation at the right level of abstraction. A plant manager approving a maintenance recommendation does not need to understand gradient descent. The manager does need to know which sensor readings drove the recommendation, how confident the model is, and what conditions would lead it to a different conclusion.

Two explainability techniques are mature enough for production use. SHAP (Shapley Additive exPlanations) values attribute a model's output to its input features, showing which drove a given prediction. LIME (Local Interpretable Model-agnostic Explanations) builds a locally accurate approximation of the model's behavior near a specific input. The Dayton recommendation card included a plain-language explanation line from the SHAP analysis that identified the two or three contributing signals. Explainability is also a governance tool: when an audit requires demonstrating the basis for a decision influenced by an AI recommendation, a dated source-attributed explanation artifact is the difference between a clean response and a protracted investigation. Aisha's team built this requirement into the data retention policy.

Diagram 6.8 - Model Risk Management Lifecycle for Twin AI

Model Risk Management Lifecycle for Twin AI

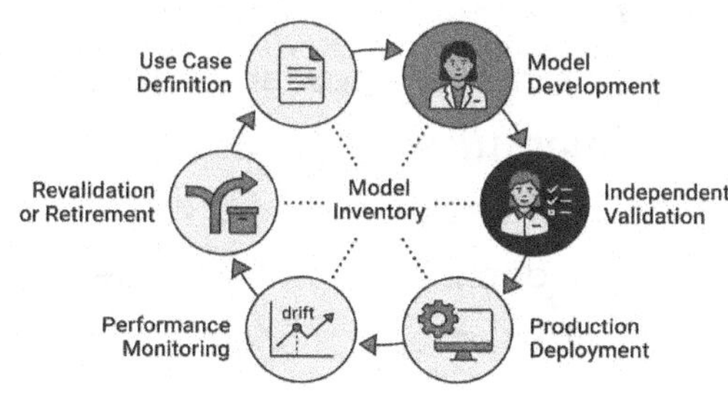

Model risk management lifecycle is saine the vi and creataticprtorming for Twin AI.

7.9 What Managers Must Approve Before Extending Autonomy

The Autonomy Approval Gate

Renata Vance's question at the March steering meeting deserves a structured answer: at what point does the system stop recommending and start acting on its own, and who approved that transition? The answer must always trace to a specific governance decision, documented with rationale and conditions. Autonomy creep, the gradual expansion of AI authority through incremental technical changes that individually seem innocuous, is a real pattern in enterprise AI programs. The defense is an explicit autonomy approval gate: defined criteria that must be satisfied, and defined people who must sign off, before any AI component transitions from advisory to execution authority.

Gate criteria include: validated model performance across the specific conditions the system will encounter (not just aggregate test metrics), explicitly enumerated authority limits, a tested and accessible override mechanism, a fallback plan for when the AI component is suspended, documented operator competency, and a monitoring plan with defined escalation triggers. None should be waived for urgency or competitive pressure. At NorthArc, the gate is a signed approval document that includes the scope description, technical validation evidence, Aisha's team's risk assessment, the business manager's operational readiness attestation, and the CDO's signature. Safety-critical capabilities additionally require the Chief Risk and Compliance Officer's signature. A well-prepared proposal moves through in two to three weeks.

Building Operational Trust Over Time

Trust in an AI system's authority to act builds through operating history: behavior documented across conditions, edge cases, and stress events. Programs that skip shadow-mode and limited-deployment phases systematically underinvest in this foundation. The evidence portfolio that supports authority expansion includes the model validation record, the shadow-mode log comparing AI recommendations to human operator decisions and their outcomes, post-deployment performance reports, and the incident record for unexpected events.

NorthArc's ninety-day shadow-mode trial for the Dayton maintenance work-order agent generated four hundred and twelve recommendations. Tomás Reyes's team acted on 328 and deferred the rest; outcomes were documented throughout. The steering committee reviewed a one-page summary. Renata's approval for production deployment came with three conditions: a written override procedure posted in the control room, a monthly performance report to the steering committee for the first six months, and automatic escalation to human review if recommendation acceptance rates fell below seventy percent. Those conditions were governance, not skepticism.

Diagram 6.9 - Autonomy Approval Gate: Evidence and Sign-Off Requirements

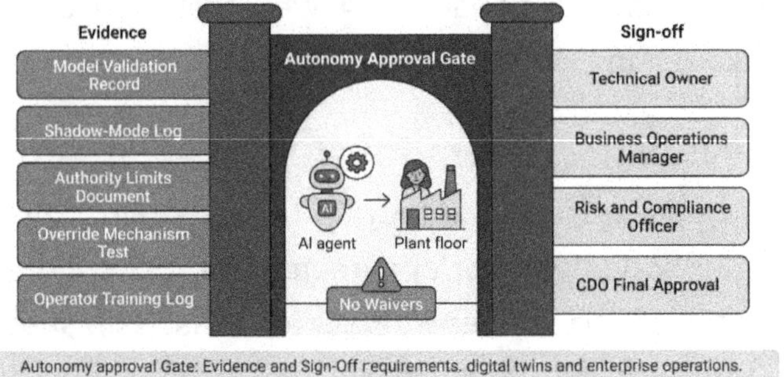

Autonomy approval Gate: Evidence and Sign-Off requirements. digital twins and enterprise operations.

7.10Manager's Checklist: Evaluating AI Overlays on Your Digital Twin

Before approving or expanding any AI capability operating over a digital twin, confirm the following:

- Stage clarity: Each AI component's current maturity stage is identified, and the criteria for authorizing the next stage are documented.
- Model inventory current: Every AI model in production or shadow mode is listed with use-case description, validation status, performance metric, model owner, and revalidation schedule.
- Surrogate model envelope documented: The valid operating envelope for each surrogate model is documented, and a monitoring process triggers review when conditions approach that boundary.
- Anomaly detection thresholds are policy decisions: Thresholds for safety-critical assets are in a formal decision register with business rationale and approving

manager's name; changes require sign-off from both operations and risk.

- Scenario simulation is scoped: Simulation tools are scoped to specific decision types with defined confidence standards before outputs influence consequential decisions.
- LLM copilot is retrieval-grounded: Natural-language interfaces over twin context retrieve from validated twin data; all queries and responses are logged with source attribution.
- RL policies have transfer protocols: Any RL policy trained against the twin simulation has a documented transfer protocol covering validated operating conditions, mandatory human override conditions, and the fallback control strategy.
- Agentic authority is explicitly enumerated: Permitted and prohibited actions are specifically listed; authority is not defined by optimization objective alone.
- Human-in-the-loop boundaries are written: Decision categories that must remain with a human regardless of model confidence are documented and communicated to all operators and system owners.
- Autonomy approval gate has been used: Every transition from advisory to execution authority has been processed through the formal gate with signed evidence from the technical owner, operations manager, risk officer, and CDO.
- Override mechanisms are tested and accessible: Human override is tested at least quarterly; operators

demonstrate competency; atrophied override capability is treated as a program-level risk.

- Explainability artifacts are retained: For any AI recommendation that influenced a consequential decision, an explanation artifact with source attribution is retained for the same period as the decision record.
- Model revalidation is budgeted: The model inventory includes a revalidation schedule and the program budget funds it; deferred revalidation is escalated to the model risk owner, not silently carried forward.

7.11Takeaway

The AI capabilities that make a digital twin a thinking twin are real and mature enough for enterprise production deployment. Surrogate models and PINNs enable high-fidelity predictions at operational speed. Anomaly detection, when governed with thresholds treated as policy decisions, reduces downtime without drowning operators in noise. Reinforcement learning, trained against the twin before deployment, enables control optimization that outperforms human-designed policies in well-defined domains. LLMs copilot information retrieval and reduce the expertise barrier to accessing twin intelligence. Agentic workflows, narrowly scoped and rigorously governed, can move from recommendation to action faster than human approval loops allow.

What none of these capabilities changes is the accountability structure that governs operational decisions. The question Renata asked about when the system stops recommending and starts acting is not one any vendor can answer on an enterprise's behalf. It is a governance decision that belongs to the leadership team, informed by validated evidence, supported by explicit authority-limit policies, and backed by the willingness to enforce those policies when commercial pressure pushes toward faster autonomy than the evidence supports. The framework in this chapter provides the structure to stay on the right side of that boundary.

Chapter 7 addresses the organizational and change management dimensions: building the cross-functional team that can operate a thinking twin at scale, sustaining operator trust through system changes, and structuring the governance bodies that keep strategic ambition and operational reality aligned over a multi-year program horizon.

8 Ownership, Access, and Accountability: Governing Digital Twins

8.1 The Near-Miss That Changed Everything

A third-party maintenance contractor working under a rotating services agreement had spent the better part of a week pulling real-time telemetry feeds from NorthArc's compressor-station digital twin. The data included pressure-cycling signatures, thermal-gradient histories, and valve-actuation sequences across six active facilities at NorthArc Energy Services. The contractor's stated purpose was a scheduled vibration analysis on two turbines. The data accessed covered all six stations and spanned eighteen months. No one had flagged the discrepancy because no one had configured the system to flag it.

Aisha Bramwell, NorthArc's Chief Risk and Compliance Officer, was briefed within the hour. Her team confirmed that the contractor's access token had been provisioned with analyst-level permissions across the entire Energy Services data domain, not the two-asset scope specified in the original work order. The token had been live for eleven days. The access log existed but had not been reviewed in the absence of a defined review cadence. No data had been exfiltrated, as far as anyone could tell, but NorthArc had no way to prove that with certainty, and that uncertainty was itself the problem.

The near-miss exposed three governance gaps simultaneously: an ownership gap, an access architecture gap, and an audit gap. Aisha's governance review ran for three weeks and produced structural changes that now anchor NorthArc's digital twin program. The gaps NorthArc discovered are not unique to NorthArc. They are the default state of most enterprise digital twin programs that have grown quickly without a well-defined governance design.

Diagram 7.1 - Governance Near-Miss: Three Gaps Exposed

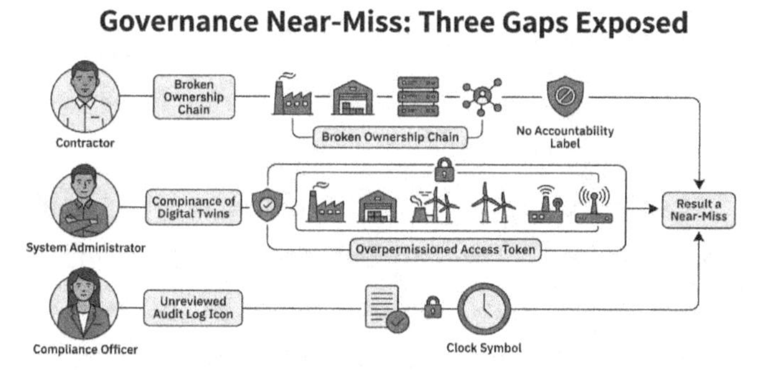

Governance near-miss simultanous asout digital twins in at digital twins of business and enterprise operations.

A digital twin program at production scale will almost certainly reveal governance gaps that were invisible during the pilot. The twin generates telemetry, derived insights, simulation outputs, and control recommendations, each with a different risk profile and regulatory obligation. Governance is not a compliance checkbox to be checked after the technology is running. It is the set of structural decisions that determine who owns what, who can see what, who can act on what, and who is accountable when something goes wrong.

This chapter covers the full governance architecture that enterprise leaders must establish: ownership models and accountability assignment, role-based access tiers, zero trust applied to twin data and control surfaces, audit logging and immutable trails, model governance under recognized frameworks, sector-specific regulatory obligations, incident response runbooks for twin compromise, and vendor liability and data residency requirements.

8.2 Ownership Models: Who Owns the Twin?

The Ownership Question Is Not Rhetorical

Before the incident, the answer to 'who owns the compressor-station twin?' would have produced three different answers depending on whom you asked. Energy Services operations claimed data ownership. Dr. Mei Lin Park's enterprise architecture group held platform ownership. Aisha's compliance team found ownership unclear, which was itself the answer. Each perspective was correct within its own frame, and the absence of a single authoritative assignment is precisely what allowed the contractor's access situation to persist.

Ownership in a digital twin context decomposes into four distinct dimensions: data ownership (raw sensor feeds and historian records), platform ownership (software environment and integration pipelines), model ownership (simulation and predictive models), and output ownership (insights, alerts, and recommendations). A business unit may legitimately claim data ownership while IT holds platform ownership and a shared analytics function owns the models. Each dimension requires a named accountable party.

Three Structural Ownership Models

Enterprise digital twin programs typically settle into one of three structural ownership models, each with distinct governance implications.

The business unit ownership model places the sponsoring operational unit in the accountable role for the twins' data, outputs, and access decisions. This model works well when the twin is scoped to a single facility and the business unit has sufficient technical depth to discharge the accountability. The risk is fragmentation: when multiple business units independently own twins sharing underlying infrastructure, governance consistency breaks down. NorthArc Manufacturing and NorthArc Energy Services ran separate twin environments under this model before the incident, and an IT administrator provisioned the contractor's token with no visibility into the Energy Services work order scope.

The IT and shared service model centralizes platform accountability within IT or a center of excellence, while business units retain domain authority. It requires a clear boundary between what IT owns (infrastructure, integration, access provisioning) and what business units own (data quality, model parameters, output interpretation). The federated shared ownership model distributes accountability across a governance council representing operations, IT, compliance, and data management. It scales better across multi-domain enterprises but requires sustained investment in governance.

Diagram 7.2 - Three Structural Twin Ownership Models

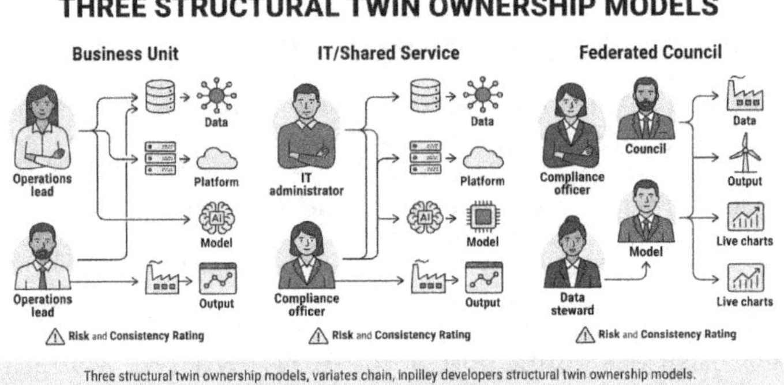

The RACI for Twin Lifecycle

Ownership assignment is most effective when formalized through an RACI matrix spanning the full twin lifecycle: data acquisition, model development and validation, deployment, access provisioning and revocation, and decommissioning. The RACI must name specific individuals, not rotating titles. NorthArc's review found the accountable role for contractor access provisioning listed as 'IT Manager' in the general security policy, with no specific individual named for the twin program.

The RACI must be reviewed at each major milestone and updated when organizational changes affect named accountable parties. A RACI that exists only in a kickoff slide deck is a historical record, not a governance artifact.

Aisha's team produced an RACI that assigns accountability for data acquisition to the business unit data steward, platform operations to Dr. Mei Lin Park's enterprise architecture function, model governance to Dorian Whitlow's AI team, and access provisioning to a named security operations lead. Aisha's compliance function holds the Accountable role for audit log review on a thirty-day cadence, a direct response to the gap the incident exposed.

8.3 Role-Based Access Tiers

Defining the Access Tier Architecture

A digital twin platform exposes fundamentally different kinds of value to different user classes. An operator needs real-time status and alarm feeds. An analyst needs historical trend data and access to simulations. A vendor performing maintenance needs narrow, time-bounded access to specific asset parameters. An executive needs aggregated KPI dashboards, not direct access to raw telemetry. Treating all of these users as equivalent in the access model is a security and governance failure.

NorthArc's architecture defines four access tiers. The operator tier covers plant-floor personnel and control room staff. Access is limited to real-time monitoring dashboards, alarm acknowledgment, and runbook execution within their assigned facility: no historical export and no simulation access. An MFA is required for any session outside the local control network.

The analyst tier includes reliability engineers and data scientists who need access to historical data, simulation capabilities, and model configuration. Data-scoping rules prevent access to facilities or asset classes outside the analyst's business unit unless a formal cross-domain access request is approved. Simulation runs that generate prescriptive recommendations for live operations require secondary approval before the recommendation surfaces to an operator.

Diagram 7.3 - Four-Tier Role-Based Access Architecture

Four-Tier Role-Based Access Architecture

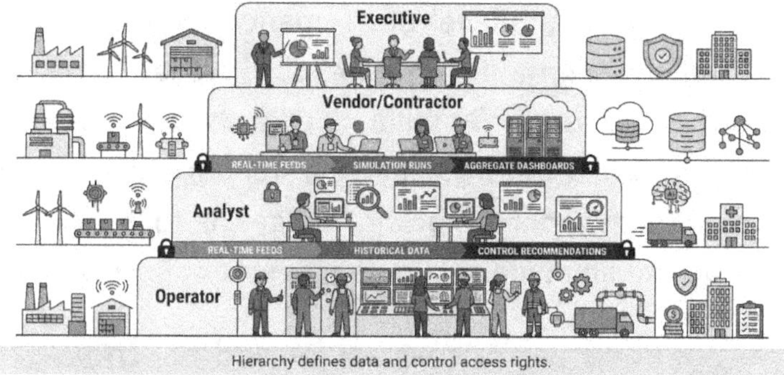

Hierarchy defines data and control access rights.

Vendor and Contractor Access

The vendor and contractor tier is where most access governance failures originate. Vendors need real access to do legitimate work. Still, their access scope is often defined by a work order that does not map cleanly to the platform's permission model, and their access lifetime is bounded by the engagement rather than by an employment relationship.

NorthArc's revised vendor access protocol requires four elements for any token issuance. First, a written access scope document signed by the business unit data steward specifying exact assets, data types, and time window. Second, a technical review confirming the requested scope maps to the minimum permission set. Third, automatic token expiration set to the work order end date plus a forty-eight-hour grace period, with no extension without secondary approval. Fourth, a real-time alerting rule flagging any out-of-scope access event within four hours. Each element directly addresses a specific component of the near-miss.

Vendor access to the executive tier does not exist. If a vendor needs program-level performance data, it should be delivered through a governed export process, not live dashboard access. Priscilla Okonkwo's vendor management practice now includes a standard contractual clause requiring vendors to acknowledge the four-element protocol before any twin-related engagement begins.

Executive Access and Need-to-Know Aggregation

The executive tier presents a different governance challenge. Senior leaders need visibility into twin-derived performance data but should not have access to raw operational telemetry. The governance question is not whether to restrict executive access but how to design the aggregation layer so that executive dashboards cannot be reverse-engineered to reveal sensitive asset-level data. In regulated environments, an executive who inadvertently receives telemetry that triggers a disclosure obligation has created a compliance problem that an aggregated KPI dashboard would have avoided.

Renata Vance's quarterly steering committee reports draw from an executive dashboard layer that applies automatic aggregation to sensor-derived metrics, suppresses facility-level breakdowns, and carries a data classification label. The dashboard is maintained by Dr. Mei Lin Park's architecture team with input from Aisha's compliance function. Any new metric requires a classification review before it is added.

8.4 Security Architecture: Zero Trust Applied to Digital Twins

Why Perimeter Security Is Insufficient

The traditional perimeter security model assumes that threats originate outside the network and that entities inside can be trusted. A digital twin environment invalidates both assumptions. The twin integrates data from OT networks, enterprise IT systems, cloud platforms, and external feeds. The OT-IT boundary, historically a meaningful perimeter, is now a managed interface that data crosses continuously. An insider threat, a compromised contractor credential, or a misconfigured integration can expose the twin's control surface entirely from inside the perimeter.

Zero trust applied to a digital twin environment means treating every access request as untrusted until verified, regardless of network origin. Verification is continuous: behavioral signals are monitored, and anomalies trigger step-up authentication or session termination. Authorization is enforced at the data layer: a user cannot access data outside their provisioned scope, even from inside the trusted internal network.

Control Surface Protection

The most consequential security distinction in a digital twin program is the boundary between the monitoring surface and the control surface. A twin that only observes and reports has a manageable security profile: a breach exposes sensitive data but does not affect physical operations. A twin that issues control recommendations or triggers automated responses has a control surface, and a successful attack there is an operational incident with potential safety consequences.

NorthArc's compressor-station twin could push recommended set-point adjustments to the DCS operator interface. The contractor did not have control surface access, but the same overly permissive provisioning model could have applied to a future engagement that did have it. Aisha's review established a categorical rule: any control surface access requires at least two named approvers, mandatory session recording, and a real-time alert to the operations duty manager upon session initiation.

Diagram 7.4 - Zero Trust Architecture for Digital Twin Environments

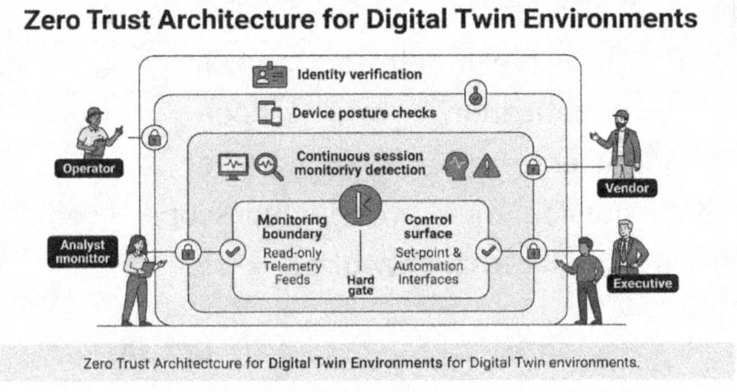

Zero Trust Architectcure for Digital Twin Environments for Digital Twin environments.

Session recording on control surface access is not optional. Every action, user identity, timestamp, and before-and-after value must be captured in an immutable record. When unexpected equipment behavior is observed, the first question is whether a human action through the twin preceded it. Without session recording, that question cannot be answered with confidence.

OT-IT Convergence Risk

OT networks were designed for availability and safety, not for the authentication and encryption standards that enterprise IT security requires. Many industrial control systems and edge gateways in active production cannot support modern identity protocols without hardware or firmware upgrades, which carry their own operational risk. The digital twin integration layer sits directly at this convergence point.

NorthArc's integration architecture uses a one-way data diode at the OT-to-IT boundary: sensor data flows through a hardware-enforced unidirectional gateway, and no traffic traverses that boundary in the reverse direction. Control recommendations are delivered as advisory messages on the IT side; the operator initiates any corresponding action through the standard DCS interface on the OT side. Tomás Reyes, who manages the Dayton facility's control room, explicitly supported this design: a twin that could write directly to the DCS without a human checkpoint was one he could not fully trust.

8.5 Audit Logging and Immutable Trails

What Must Be Logged

Audit logging in a digital twin environment covers three event categories. Data access events record who accessed what data, from which endpoint, at what time, and for how long. In the absence of automated scope enforcement, access logs are the only mechanism for detecting over-access. The contractor incident at NorthArc was detected through log analysis, not through automated alerting, because the alerting rules had not been configured. That sequence should not be the standard operating mode.

Model execution events record when a model was run, what inputs were provided, what outputs were generated, and who triggered the run. This logging category is essential for post-incident analysis when a model's output contributed to a decision that led to unintended consequences. Configuration change events record any modification to data sources, model parameters, alert thresholds, or access permissions. Configuration changes outside a formal change management process are a frequent precursor to security incidents.

Diagram 7.5 - Audit Log Architecture: Three Event Categories

Audit Log Architecture: Three Event Categories

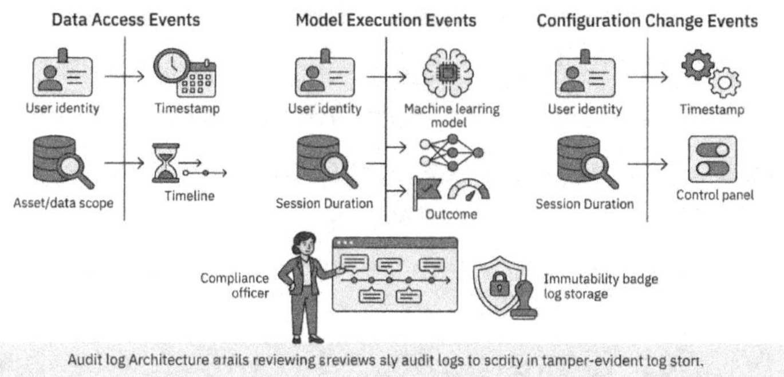

Audit log Architecture atails reviewing sreviews sly audit logs to scdity in tamper-evident log storr.

Immutability and Retention

Immutable audit logging means that log records, once written, cannot be altered or deleted by any user or process, including system administrators. This is implemented through write-once storage with a separate administrative control boundary, or through cryptographic chaining that makes modification detectable. In regulated environments, immutability is often a compliance requirement rather than a design choice.

Retention periods should be defined based on the applicable regulatory minimum, the organization's incident response timeline, and vendor agreement requirements. Aisha's team set the audit log retention to a multiple of the regulatory minimum to accommodate a realistic examination timeline after an incident. Technically, retained logs that are not operationally searchable do not serve their governance purpose: retention must be paired with an index and query capability that surfaces a specific access event within a defined response window.

Aisha established a 30-day review cadence for contractor access logs and a 90-day cadence for internal user logs. An automated anomaly detection layer flags sessions that exceed defined scope or duration thresholds. Human review is not optional even when automation finds nothing: confirming the automated layer's completeness is itself a governance control.

8.6 Model Governance: Frameworks and Practices

Why Model Governance Is Inseparable from Twin Governance

A digital twin at any maturity level beyond the descriptive stage operates models. Diagnostic twins apply anomaly detection to telemetry streams. Predictive twins run forecasting models over time-series data. Prescriptive twins use optimization models to generate recommended actions. Each model can drift, fail, or produce systematically biased outputs in ways that are not immediately visible to the operators acting on them. A model governance failure in a digital twin program is an operational decision quality problem, not an IT problem.

Dorian Whitlow's team learned this directly. A predictive maintenance model trained on a narrow operating range began understating failure risk when two stations were pushed to higher throughput during a seasonal demand spike. The model continued to generate outputs confidently even as its input distribution shifted outside its training envelope. A routine model performance review caught the drift before any maintenance decision was affected, but only because the review cadence had been established and was being followed.

NIST AI RMF, ISO/IEC 42001, and Model Cards

The NIST AI Risk Management Framework provides a four-function structure (Govern, Map, Measure, Manage) that applies directly to the models within a digital twin program. Govern covers policies and accountability structures. The map identifies the risks a specific model presents in its operational context. Measure monitors model performance against defined metrics. Manage cover response actions when a model's risk profile changes. Dorian's team uses the RMF as the organizing structure for model lifecycle documentation, producing a model card at deployment that carries forward through every subsequent review cycle.

ISO/IEC 42001 extends the governance structure to the organizational level. Its most operationally relevant elements are documented AI policy, defined roles and responsibilities for AI governance, and records of AI-related decisions and risk assessments. These requirements map naturally onto the RACI structure Aisha's team established and reinforce treating model governance as a program-level discipline, not a task for individual developers.

Model cards are the practical artifact that makes model governance operational. A model card for a twin prediction or optimization model should document: intended use case and its limits, training data characteristics and operational range, performance metrics at deployment, and the monitoring thresholds that trigger review, known failure modes, approval signatures for deployment and parameter changes, and the escalation path when metrics fall outside bounds. A model card produced once and never updated is a deployment documentation artifact, not a governance artifact. The governance value comes from the review cycle.

SR 11-7 Style Model Risk Management

The Federal Reserve's SR 11-7 supervisory guidance was written for financial services. Still, its three-part structure (development review, ongoing monitoring, independent validation) applies to any environment where model outputs influence consequential decisions. NorthArc is not a bank, but a predictive maintenance model that systematically understates failure risk, and its financial and safety consequences are structurally similar to the credit risk model failures SR 11-7 was written to prevent.

The most practically applicable element of SR 11-7 discipline is independent model validation: the function that reviews a model's methodology, data, and performance should be organizationally separate from the function that developed it. At NorthArc, Dorian's team develops predictive models, and Dr. Mei Lin Park's architecture team performs validation reviews, with input from Aisha's compliance function, for models in regulated asset classes.

Diagram 7.6 - Model Governance Lifecycle: Development to Decommission

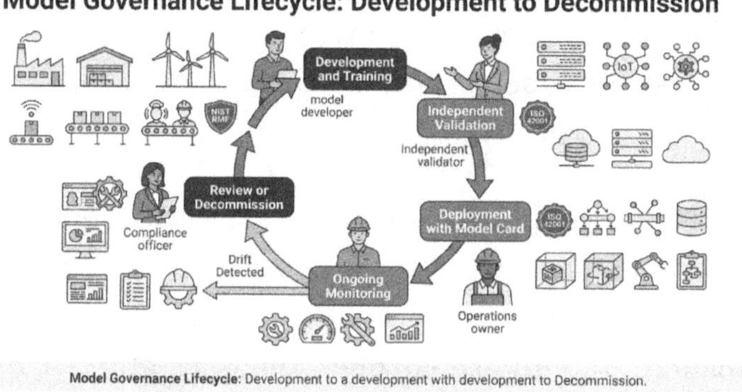

Model Governance Lifecycle: Development to a development with development to Decommission.

8.7 Regulated Environments: Sector-Specific Obligations

Utilities and Critical Infrastructure

Digital twin deployments that touch grid-connected assets impose obligations beyond general enterprise security and data governance. NERC CIP standards define specific requirements for electronic security perimeters, physical security of cyber assets, and access management for systems that could affect the reliable operation of the bulk electric system. A twin that receives data from or sends control signals to bulk electric system assets is likely subject to CIP requirements depending on its classification.

Utilities should conduct a formal scope assessment with qualified cybersecurity counsel before deployment to determine which CIP standards apply. Key questions include whether the twins' integration layer constitutes an Electronic Security Perimeter, whether personnel require CIP background screening, and whether data retention and incident reporting must align with CIP-008. These require a cross-functional assessment involving operations, legal, compliance, and cybersecurity expertise.

Healthcare

Healthcare organizations face a layered regulatory environment. HIPAA's Privacy and Security Rules apply if the twin processes individually identifiable health information. The threshold question is whether a data element or a combination of data elements that enables re-identification constitutes protected health information under HIPAA's definition. A facility's twin-tracking room occupancy by department at the aggregate level may not trigger HIPAA. The same twin, redesigned to track individual patient locations in real time, almost certainly does.

Healthcare organizations working with federal payers may also face FedRAMP requirements for cloud-hosted platform components. FedRAMP authorization must be planned at the architecture stage, not discovered after deployment. Medical device twins that influence device configuration or clinical decision-making may be subject to FDA oversight. Governance teams in healthcare should map every data flow against HIPAA, FedRAMP, and applicable FDA guidance before finalizing the platform architecture.

Food, Defense, and Other Regulated Sectors

Food manufacturing operations must consider how twin outputs interact with FSMA record-keeping requirements and HACCP documentation obligations. If the twin monitors critical control points and generate automated alerts, those alerts and the underlying telemetry may constitute records that the FDA can examine during an inspection. Ensuring that audit logs and model outputs are organized and searchable in a format supporting FDA examination is a governance requirement, not a development aspiration.

Defense contractors and government agencies must comply with the NIST Risk Management Framework, DISA STIGs, and the Authorization to Operate (ATO) process for any twin connecting to a classified network. The ATO must be integrated into the program plan from the beginning; retrofitting authorization documentation onto an already-deployed system is significantly more costly than building that discipline into the development lifecycle.

Diagram 7.7 - Regulatory Overlay by Sector

Regulatory Overlay by Sector

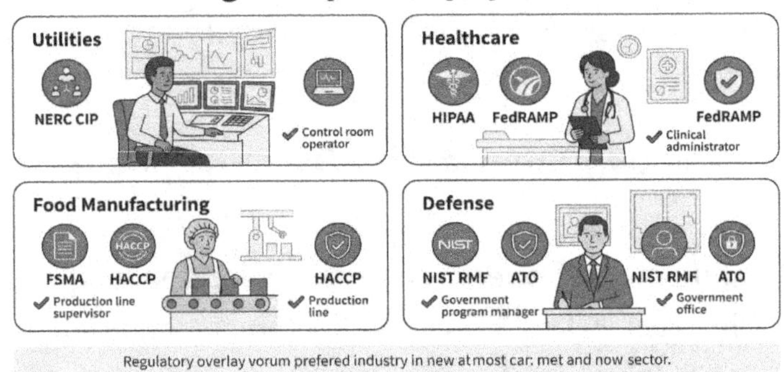

Regulatory overlay vorum prefered industry in new at most car: met and now sector.

8.8 Incident Response Runbooks for Twin Compromise

Why Twins Require Dedicated Runbooks

A digital twin compromise does not fit neatly into a standard IT incident response framework. The consequences depend on which surface was accessed. Unauthorized access to historical telemetry is a data breach with a defined regulatory notification timeline. Unauthorized modification of a model parameter is a data integrity incident with operational consequences. An unauthorized command through the control surface is an operational incident that may require immediate physical safety response. Standard IT runbooks rarely account for the third category.

NorthArc's incident response program includes three dedicated twin compromise runbooks, one per category. Each runbook defines detection triggers, immediate containment actions, the escalation chain with named individuals, evidence preservation requirements, regulatory notification timelines, operational impact assessment process, and the criteria for closing the incident and transitioning to after-action review.

Containment Actions Specific to Twin Architecture

Containment actions differ from standard server incident response because isolating the affected system may itself create operational risk. A compressor-station twin providing real-time monitoring cannot simply be taken offline during a security investigation without degrading the control room's situational awareness. The runbook must specify a degraded-mode operating procedure that is trained and practiced before it is needed.

For data access incidents, the immediate containment action is to revoke credentials. This requires that the access control system support immediate revocation without waiting for a token expiration cycle. Systems that use long-lived tokens without a revocation mechanism create a containment gap: unauthorized access may persist until the token expires. NorthArc's architecture review resulted in a maximum contractor token lifetime of 24 hours and a server-side revocation capability that terminates any active session immediately upon a security operations command.

Diagram 7.8 - Twin Compromise Incident Response Flow

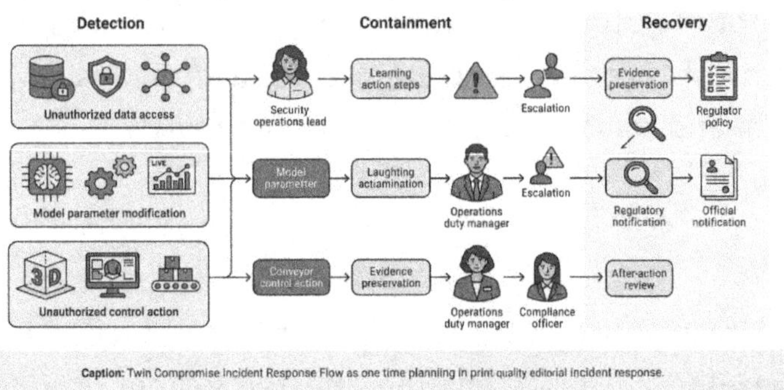

Caption: Twin Compromise Incident Response Flow as one time planniing in print quality editorial incident response.

After-action review should produce three outputs: a root cause analysis, remediation actions with owners and completion dates, and an updated runbook. Incident response capability is built through exercise. NorthArc runs an annual tabletop exercise of its twin-compromise runbooks and an unannounced credential revocation drill every 18 months. Tomás Reyes participates in every drill because the degraded-mode procedure requires execution on the plant floor.

8.9 Vendor Liability, Data Residency, and Contractual Governance

What Your Platform Contract Must Address

Priscilla Okonkwo's vendor management practice evolved significantly after the governance review. The original twin platform contracts covered commercial terms but left critical governance dimensions absent or inadequate, requiring renegotiation in two cases and execution of addenda in three others.

Every twin platform contract should address five governance dimensions. Data ownership and permitted use: the enterprise retains ownership of all raw sensor data, derived outputs, and model configurations. Define whether the vendor may use aggregated data to train platform models. Breach notification: the vendor notification timeline must be shorter than the applicable regulatory deadline so the enterprise can conduct its own assessment before the external clock starts.

Data residency specifies where data is stored and processed, a compliance requirement in any sector with data localization obligations or cross-border transfer restrictions. Vendor access to production data should require advance notice, a defined scope, and an enforceable logging obligation. Liability allocation must specify how responsibility is distributed when a platform vulnerability or vendor error contributes to an incident, including indemnification terms and liability caps.

Subprocessor and Fourth-Party Risk

Digital twin platforms rely on cloud infrastructure providers, AI inference services, time-series database vendors, and integration middleware. Each subprocessor represents a point at which twin data is shared with a party that is not directly under contract with the enterprise. In regulated environments, GDPR, HIPAA, and sector-specific regulations require enterprises to know where their data goes and to ensure downstream processors provide equivalent protections.

Priscilla's vendor questionnaire now includes a subprocessor disclosure requirement, a notification obligation for any change to the subprocessor list, and a right-to-audit provision that allows NorthArc to request evidence of compliance assessment. A vendor who resists subprocessor disclosure is providing information about its own governance maturity that should be factored into the procurement decision.

Diagram 7.9 - Contractual Governance Architecture: Vendor and Subprocessor Map

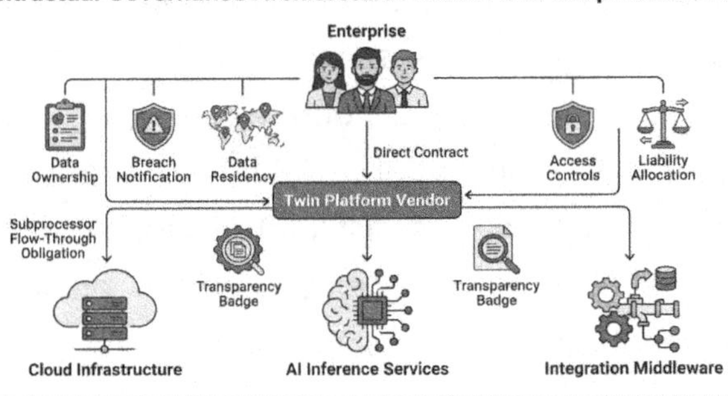

Ensuring consistent governance and compliance across vendor partnerships and their subprocessor network.

Data residency deserves particular attention in multi-cloud or hybrid deployments. A platform that stores primary data in a compliant region but processes simulation workloads in a lower-cost region may create an enterprise-unaware cross-border transfer. Requiring vendors to document the geographic data path from ingest through processing through storage is a reasonable governance requirement that should be incorporated into the initial platform evaluation.

8.10 Manager's Checklist: Governance Posture for Digital Twin Programs

Work through this checklist before deployment and revisit it at each major program milestone.

- Ownership assignment: Named individuals hold accountability for data, platform, model, and output ownership. The RACI spans the full twin lifecycle and names specific individuals, not rotating titles.
- Access tier architecture: Four access tiers are defined (operator, analyst, vendor/contractor, executive) with documented permission sets. No tier has access beyond what its operational function requires.
- Contractor access protocol: The four-element protocol (written scope document, technical review, automatic token expiration, real-time scope alerting) is in place and tested before any external party receives a token.
- Zero trust: Authentication is continuous, authorization is enforced at the data layer, and no user or system holds blanket access based on network location.
- Control surface protection: Control surface access requires a minimum of two named approvers, mandatory session recording, and a real-time alert to the operations duty manager on session initiation.

- OT-IT boundary: The integration architecture enforces a unidirectional data flow from OT to IT. No network path enables remote actuation from the IT network to OT systems without a human checkpoint.
- Audit logging: All three event categories (data access, model execution, configuration change) are captured with immutable storage, a defined retention period, and a review cadence with named reviewers.
- Model cards: Each predictive or optimization model has a current model card documenting intended use, training data scope, performance metrics, known failure modes, approval signatures, and escalation path.
- Independent model validation: Validation is performed by a function organizationally separate from the development team on a defined cadence.
- Regulatory scope assessment: A formal, documented assessment with qualified compliance counsel has determined which sector-specific requirements (NERC CIP, HIPAA, FSMA, RMF, others as applicable) apply to the twins' architecture and data flows.
- Twin compromise runbooks: Dedicated runbooks exist for data access, model integrity, and control surface incidents. Each includes a degraded-mode operating procedure. Runbooks are exercised on a defined cadence.
- Platform contract governance: Vendor contracts address data ownership, breach notification, data

residency, vendor access conditions, liability allocation, and subprocessor disclosure.

- Governance repository: A single authoritative repository holds the RACI, model cards, access policy, twin inventory, audit log retention schedule, runbooks, and contractual records, updated on a defined schedule.

8.11Takeaway

The governance of a digital twin program is not a separate workstream alongside the technical program. It is the structural decisions that determine whether the program can be trusted by the operators who act on its outputs, the executives who make investment decisions based on its insights, and the regulators who assess the organization's risk posture. Aisha Bramwell's response to the contractor near-miss was to redesign the governance architecture from the ownership layer through access tiers, the audit trail, model governance, and incident response runbooks.

These decisions do not require a large governance team. They require a manager who assigns accountability to specific individuals before problems surface and treats governance artifacts as operational tools. The near-miss revealed that three governance gaps could coexist undetected in a well-run organization. The remedy was clarity about who owned what, who could access what, and who was responsible for knowing when something had gone wrong.

A technically sophisticated but governance-deficient digital twin is an operational liability. Ownership, access, and accountability are not constraints on the program. They are the conditions that allow it to operate responsibly at scale, in regulated environments, and under the scrutiny it will eventually receive.

9 Ready or Not: Skills, Teams, Vendors, and the Operating Model

9.1 The Vendor Room

The conference room on the fourteenth floor of NorthArc's Columbus headquarters had been booked for three consecutive Thursdays. Each Thursday belonged to a different vendor finalist, and the agenda was always the same: ninety minutes for a capabilities demonstration, thirty minutes for a reference architecture walk-through, and sixty minutes of structured questioning. Priscilla Okonkwo had designed the process herself, borrowing from a procurement playbook she had used years earlier to evaluate ERP systems. The parallel was imperfect, she knew, because ERP replacement was a known problem with a catalog of measurable outcomes. Selecting a digital twin platform was like choosing a long-term engineering partner for a program whose scope would expand in ways she could not fully predict.

Hector Salinas sat beside her at the far end of the table. He had attended all three sessions, not because procurement was his domain but because his manufacturing plants were the primary deployment targets for the next phase of the program. He came with a single sheet of handwritten questions. The first read: 'What happens to our data models and configuration work if we need to switch vendors in year three?' The second read: 'Who owns the physics models we build on your platform?' He had written a third question and then crossed it out. Priscilla had seen it before he covered it: 'Can your engineers actually talk to mine?' She told him to put it back. It belonged on the list.

The three finalists represented meaningfully different strategic bets. One was a hyperscale cloud provider whose platform offered deep integration with existing IT infrastructure but whose operational technology depth was thin. Another was a specialty industrial software vendor with domain-specific simulation capabilities but a smaller ecosystem and a services organization that had stretched itself across too many accounts. The third was a system integrator-led consortium offering best-of-breed components assembled under a managed services model, which came with the strongest day-one integration capability and the least predictable long-term cost structure. None of them was clearly right. All of them had conditions that required careful negotiation. What Priscilla and Hector were learning, session by session, was that selecting a twin platform was inseparable from the organizational decisions that surrounded it: what skills NorthArc needed to build internally, how the implementation team would be structured, and how they would manage the operators whose workflows were about to change.

9.2 Skill Inventory and Gap Analysis

The Core Competency Map

A digital twin program draws on at least seven distinct competency domains, and almost no organization enters a twin initiative with all seven covered. The first step for any manager is to be explicit about what the program requires, honest about what the organization currently has, and deliberate about how each gap gets closed. Vague references to 'building digital capabilities' are not a plan; a competency map with named skills, current proficiency levels, and explicit sourcing decisions is.

The seven core domains are: data engineering, machine learning and AI, simulation modeling, operational domain expertise, control systems and OT integration, governance and security, and change management. Each plays a different role in the program, and gaps in any one of them show up as specific failure modes. Lack of data engineering depth leads to pipelines that break under production load. Missing OT integration knowledge produces twins that cannot reliably ingest sensor data from legacy equipment. Missing governance capability produces models that drift without detection. Missing change management produces tools that operators distrust and avoid. The failure mode always reflects the gap.

Diagram 8.1 - Digital Twin Competency Map

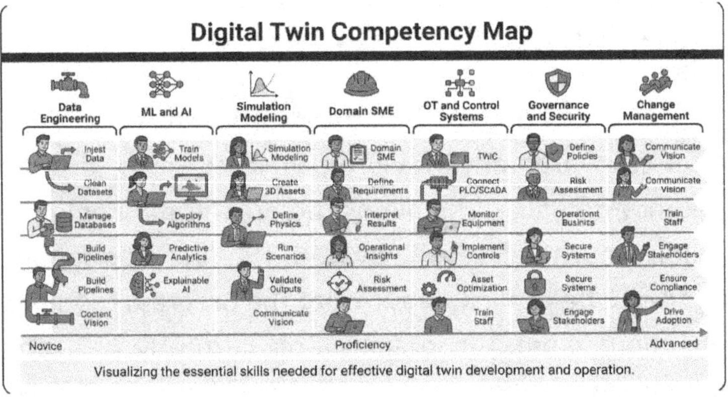

Visualizing the essential skills needed for effective digital twin development and operation.

Data engineering is frequently underestimated because program sponsors focus on the AI and simulation layers and treat data pipelines as a solved problem. In practice, OT-to-cloud pipelines for manufacturing environments involve protocol translation (OPC UA, MQTT, Modbus), edge buffering, quality filtering, and lakehouse ingestion, all of which require specialized skills that general-purpose data engineers may not have. NorthArc discovered this at Dayton, where Dr. Mei Lin Park's architecture team spent three months resolving fidelity issues from a single mismatch between the historian's timestamp resolution and the ingestion pipeline's batch interval.

Machine learning and AI in the twin context is narrower than the broad label implies. The relevant skills are anomaly detection on time-series data, physics-informed model training, surrogate model construction, and reinforcement learning for optimization loops. These differ from the skills required to build recommendation systems or natural language models, and assuming interchangeability causes misassignment. Dorian Whitlow at NorthArc spent early program months correcting this assumption, redirecting analysts assigned to the twin solely because they had done regression work in customer analytics.

Simulation Modeling and Domain Expertise

Simulation modeling sits at the intersection of engineering and software. The scarcest combination is a practitioner who can build a physics-based model of an industrial process and connect it to a live data feed. Most simulation engineers have worked with offline tools such as MATLAB, Ansys, or Aspen, and reorienting them toward real-time synchronization requires deliberate retraining, not a simple reassignment. For organizations without existing simulation capability, outsourcing initial model construction to an integrator while building internal skills in parallel is a practical path, provided the contract preserves model access and documentation rights from day one.

Domain SMEs are the category most often treated as available on demand and most often lost to competing operational priorities. A turbine engineer who understands degradation patterns well enough to validate a simulation model is typically also the engineer responsible for reactive maintenance when equipment fails. Protecting SME time requires explicit commitment from the operational business unit, not from the digital program office. Hector Salinas made this commitment for NorthArc

Manufacturing in writing, allocating 20 percent of the time of two senior engineers to the program for twelve months.

Diagram 8.2 - Skill Sourcing Decision Matrix

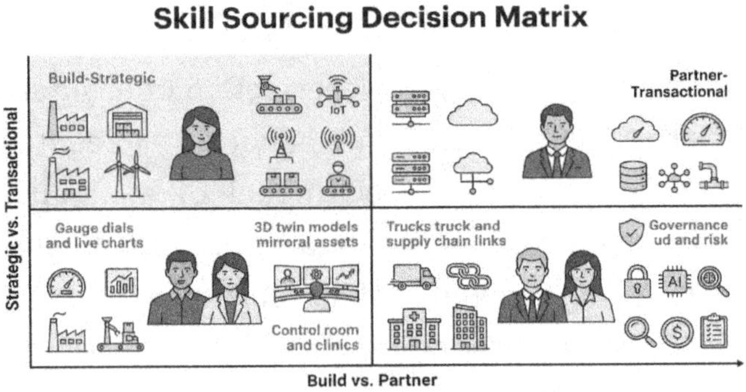

That ligit-nlinæmotions is not added a decrease to the saræ rassurement of digital twins and enterprise operations.

Governance, Security, and Change Management Skills

Governance and security competency extends beyond standard IT practices. A twin platform that knows the real-time physical state of industrial systems is a high-value adversarial target. Personnel responsible for it must understand OT security (IEC 62443 applied to OT networks) as well as AI model governance (including drift detection, bias auditing, and versioning). Aisha Bramwell's team at NorthArc had ISO 27001 covered for IT systems; the OT security dimension required a dedicated hire and a six-month skills uplift for the existing security operations team.

Change management is routinely underinvested in because it appears to be a soft cost compared to platform licensing and integration services. The practical consequence is adoption failure. Operators who distrust twin recommendations, or cannot confidently interpret the interface, revert to established routines within weeks of go-live. Effective change management for twin programs requires practitioners who understand industrial workflows, not general organizational development generalists. The best candidates come from operations consulting firms specializing in manufacturing or utilities, or from frontline supervisors within the organization who carry credibility with operators.

9.3 Team Design and Organizational Structure

Three Archetypes: CoE, Federated Guild, and Hybrid

Every organization that reaches production scale with a digital twin program has made a structural choice about where the program's capabilities live. The three dominant archetypes are: a centralized Center of Excellence (CoE), a federated guild model in which capabilities are distributed across business units with lightweight central coordination, and a hybrid model that maintains a central platform and governance function while embedding delivery capabilities within each operational unit. Each has genuine strengths and genuine failure modes, and the right choice depends on the organization's size, the diversity of its operational contexts, and the maturity of its existing data and technology functions.

The centralized CoE provides consistency, specialized depth, and economies of scale in skills development and tooling. Its characteristic failure mode is distance from operations. When the CoE is staffed by data scientists and engineers who are organizationally remote from the plant floor, requirements misalignment accumulates quietly until it surfaces as a delivery failure. The NorthArc Energy Services team experienced this with their first grid-twin prototype: the CoE had built technically correct models of the distribution network topology, but the operating parameters used in the simulation did not reflect the informal loading practices that field operators actually applied. The models were accurate in theory and wrong in practice.

Diagram 8.3 - Team Structure Archetypes for Twin Programs

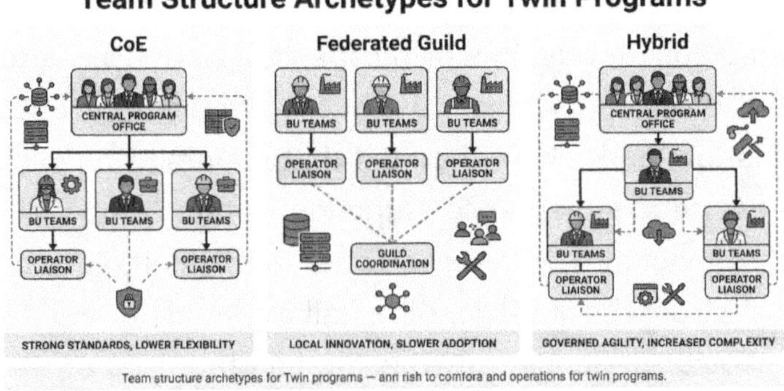

Team structure archetypes for Twin programs – ann rish to comfora and operations for twin programs.

The federated guild model distributes twin capability into business units, with each unit maintaining its own small team and a shared community of practice providing standards and knowledge exchange. Its strength is proximity to operations. Its failure mode is fragmentation: without a strong central governance function, guilds develop incompatible architectures, duplicate vendor relationships, and produce twins that cannot be integrated into an enterprise view. For organizations with highly diverse operational contexts (NorthArc spans manufacturing, energy services, and logistics with meaningfully different OT environments), some degree of federation is structurally necessary. The question is how much central coordination to maintain.

The hybrid model reflects the reality of most mature twin programs. A central platform team maintains data fabric, shared simulation infrastructure, and governance framework. Embedded delivery teams within each business unit own use-case development, interface design, and change management. The liaison role is the structural element that makes the hybrid work: without a named liaison carrying authority on both sides, the boundary between central and embedded functions becomes a coordination gap rather than a handoff.

The Liaison Role and Governance Committee Design

The liaison role is the single organizational design decision that most consistently differentiates programs that achieve operational adoption from those that do not. The liaison is not a project manager. The role requires technical credibility with the platform team, operational credibility with the business unit, and communication skills to translate requirements in both directions without distortion. At NorthArc, Tomás Reyes at the Dayton facility served informally in this capacity during the first pilot phase. When the program moved to a formal hybrid structure, Renata Vance explicitly designated the liaison role, funded it from the program budget rather than the plant's operating budget, and wrote a role definition that precisely described accountability.

Governance committee design requires balancing representation (to build legitimacy) with decision-making efficiency. A steering committee that meets monthly with four to six named decision-makers, supported by a technical advisory working group that meets bi-weekly, is workable at NorthArc's scale. The design should specify: who approves platform architecture changes, who approves new use-case commitments, who owns the vendor escalation path, and who can pause program activity if a safety concern arises.

Diagram 8.4 - Twin Program Governance Committee Structure

Twin Program Governance Committee Structure

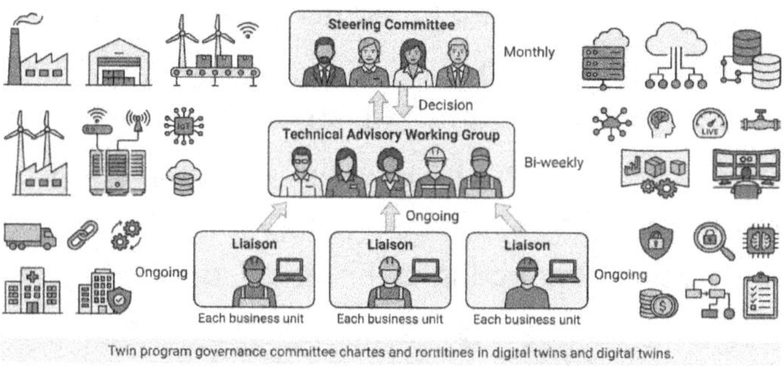

Twin program governance committee chartes and rontitines in digital twins and digital twins.

9.4 Vendor Evaluation: Selecting Partners for a Program That Will Evolve

The Platform Landscape

The digital twin platform market presents a more diverse competitive landscape than most other enterprise software categories. Vendors arrive from different directions: hyperscale cloud providers have entered through IoT connectivity and data services, industrial software incumbents have extended their engineering simulation tools to connected platforms, specialized twin vendors have built from the ground up for specific industry use cases, and system integrators have assembled best-of-breed stacks under managed services models. Understanding each vendor's strategic origin helps predict where their platform will be strongest and where integration overhead will be highest.

Microsoft Azure Digital Twins and AWS IoT TwinMaker approach the problem from the IT infrastructure and data services direction, offering strong integration with existing cloud workloads and developer ecosystems familiar to enterprise IT teams. Simulation depth is more limited compared to specialized industrial tools, and OT integration typically requires additional middleware or third-party connectors. For organizations where the primary value is connecting operational data to existing analytics investments, these platforms offer a low-friction starting point.

NVIDIA Omniverse approaches the problem from the 3D simulation and visualization direction, providing high-fidelity rendering and physics simulation that is particularly strong for manufacturing environments where spatial layout, robot coordination, and visual inspection matter. Siemens Xcelerator and the broader Siemens Industrial Operations X portfolio provide deep integration with Siemens control systems and automation hardware, making them a natural fit for brownfield environments with significant Siemens equipment installed. Ansys Twin Builder is oriented toward simulation-first programs where the physics model is the primary asset. GE Digital's Predix and SmartSignal lineage is strongest in asset performance management and predictive maintenance for rotating equipment. Bentley iTwin focuses on infrastructure and civil engineering assets, providing strong capabilities for utilities, transportation, and facilities programs. Dassault Systèmes 3DEXPERIENCE integrates PLM, simulation, and manufacturing operations management into a unified platform that suits organizations where the twin bridges of design and operations converge.

Diagram 8.5 - Digital Twin Platform Landscape

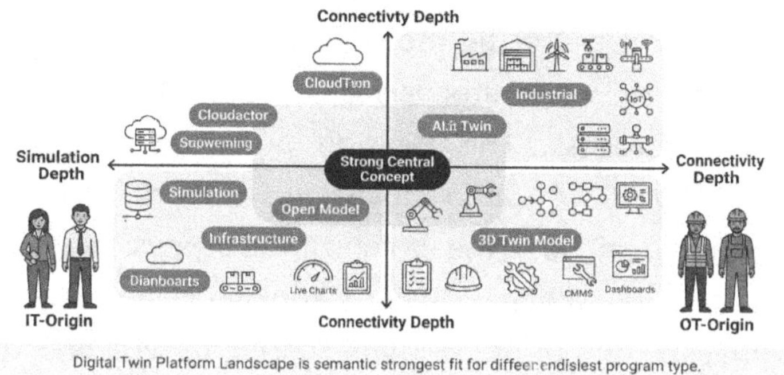

Digital Twin Platform Landscape is semantic strongest fit for diffeer endislest program type.

Build vs. Buy vs. Partner

The build vs. buy decision for digital twin capability is rarely binary at the enterprise scale. The practical question is which layers of the stack the organization should own, which should be licensed from a platform vendor, and which should be delivered as a managed service by an integrator. A useful framework separates the stack into four layers: physical model and data acquisition (typically proprietary and must be owned), integration and data normalization (often partially owned, partially licensed), simulation and analytics (typically licensed or partnered), and user experience and workflow integration (must be owned because it encodes operational knowledge). Organizations that try to build all four layers internally underestimate complexity and overestimate their software engineering capacity. Organizations that outsource all four layers lose the organizational capability to sustain, evolve, or migrate the program.

A healthy vendor partnership in this model means the vendor owns the platform infrastructure and the core simulation engine. In contrast, the customer organization owns the domain-specific model configuration, the use-case logic, the operator-facing interfaces, and the governance process. This division preserves the most important form of program value: the accumulated understanding of how the organization's physical assets actually behave. When Priscilla Okonkwo evaluated the three finalists, one of her primary scoring criteria was the vendor's policy on model artifacts: specifically, whether configuration files, physics model parameters, and trained ML components were portable, documented, and exportable in open formats.

Diagram 8.6 - Build-Buy-Partner Stack Allocation

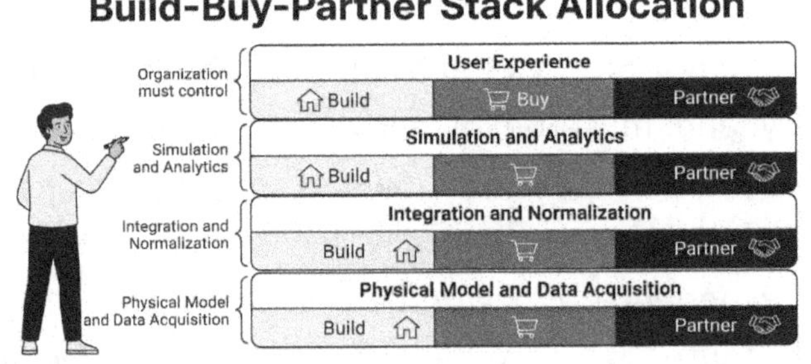

Evaluation Criteria: What Matters Beyond the Demo

Vendor demonstrations are calibrated to show platform strengths on curated datasets. A structured evaluation process uses the demo as a starting point and tests against the conditions that predict long-term success. The criteria that matter most: platform breadth and roadmap credibility, openness and data portability, ecosystem health and partner availability, services capacity, and lock-in risk.

Platform breadth and roadmap credibility should be evaluated against anticipated program evolution, not just current requirements. Priscilla Okonkwo's team requested three reference conversations for each finalist: one at a comparable maturity level, one from a customer who had expanded beyond the initial scope, and one from a customer who had worked through a significant implementation challenge. The last category was the most informative.

Openness and data portability are consistently underweighted in initial evaluations and most consequential when the program encounters problems. The key questions: in what format are simulation model configurations stored, can they be exported to other tools, are trained ML components portable (ONNX, PMML, or equivalent), and do portability provisions survive contract termination? Vendors with proprietary model formats and no export path create a dependency that makes migration prohibitively expensive. This must be priced into the long-term cost model.

9.5 Contract Terms Every Manager Must Negotiate

Data Rights, Model Ownership, and IP

Contract negotiation for digital twin platforms requires managers to engage with technical details that are easy to defer to legal or procurement staff, only to regret it when they become operational constraints. Three categories demand the operational manager's direct attention: data rights, model ownership, and IP produced during implementation.

Data rights provisions determine who can access, analyze, aggregate, and resell operational data flowing through the platform. Many vendors include clauses permitting the use of anonymized customer data for model training. The key negotiating position is specificity: require the contract to name exactly what data categories are covered, what processing is permitted, whether data is shared with third parties, and how data use rights terminate when the contract ends. Aisha Bramwell flagged two standard vendor agreements in which data retention rights survived termination for 24 months, with no obligation to delete.

Model ownership addresses a complex question: when a vendor's platform is used to train a predictive model with the customer's data and engineering expertise, who owns the resulting model? Vendor defaults vary. Some treat models as derivative works and claim co-ownership rights. The most defensible contract position is explicit customer ownership of: all trained model artifacts and parameters, all simulation configurations created during the engagement, and all documentation produced by the implementation team. This requires negotiation and is not always fully achievable, but the negotiation process itself reveals how the vendor conceptualizes the customer relationship.

Diagram 8.7 - Contract Terms Negotiation Checklist

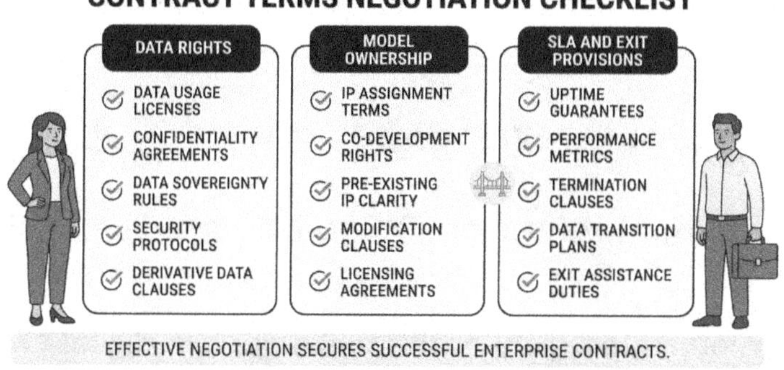

SLAs, Exit Provisions, and Lock-In Mitigation

SLAs must cover data pipeline latency, not just platform uptime. A 99.9 percent uptime guarantee that places no obligation on ingestion latency creates a legal remedy for outages while leaving the most common failure mode (stale data with no warning) unaddressed. Negotiate SLAs specifying: maximum sensor-to-twin latency under normal and high-ingestion conditions, notification thresholds when latency is breached, and defined remediation timelines.

Exit provisions are most often left to boilerplate and are most consequential when the relationship ends. They should specify: a defined export format and timeline (60 to 90 days post-termination), a transition services period, provisions for exporting trained model artifacts and configuration files, and a non-interference clause preventing the vendor from revoking API access during a contested dispute. Priscilla Okonkwo treated exit provisions as a test of vendor confidence: a vendor who resisted clean exit terms was signaling concern about retaining customers on merit.

9.6 System Integrator Selection

What Integrators Do and Where They Fall Short

System integrators play a different role in a digital twin program than they do in traditional ERP or infrastructure deployments. The value proposition of an integrator in a twin program is not primarily code delivery; it is the combination of platform-specific implementation experience, OT-IT integration methodology, and the accumulated pattern library from previous implementations that allows them to anticipate and solve problems that a first-time program would encounter as surprises. An integrator who has connected a historian to an Azure Digital Twins instance across 20 previous engagements has encountered and resolved edge cases that would otherwise have consumed months of first-time implementation.

For system integrators, platform breadth matters less than depth of experience with the specific platform the organization has selected. References from similar industries and complexity profiles are more valuable than general capability statements. The composition of the proposed delivery team is more predictive of outcome than the firm's overall reputation: request the engineers' CVs who would actually be on the engagement, verify certifications, and confirm availability before contracting. Integrators routinely win business with senior staff and delegate delivery to less experienced teams. Hector Salinas's third question was the right one.

Diagram 8.8 - System Integrator Evaluation Scorecard

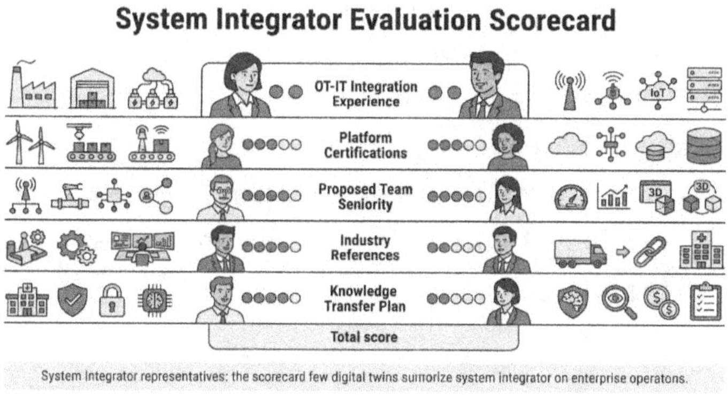

Knowledge transfer obligations are the most consistently underspecified element of integrator contracts. An integrator who builds a twin and retains all the tacit implementation knowledge creates perpetual dependency. Contracts should specify documented architecture decisions with rationale, annotated code repositories with naming and commenting standards, runbooks for all operational procedures, and a formal knowledge transfer period of at least four weeks at program conclusion, during which internal staff shadow the integrator's team on all maintenance and modification tasks. The knowledge transfer period should be a contractual milestone with an acceptance criterion, not an informal courtesy at the close of engagement.

Fee structures require care because the scope typically evolves as the program learns. Time-and-materials provides flexibility but creates incentive misalignment when velocity matters. Fixed-scope provides cost predictability but puts pressure on minimum viable delivery. A phased approach using fixed scope for well-defined workstreams (pipeline build, platform configuration, baseline model training) and time-and-materials for exploratory workstreams (new use-case development, novel integrations) aligns incentives to the nature of the work.

9.7 Change Management for the Operator Workforce

Identifying and Engaging Operators Whose Workflows Will Change

The operator workforce is the population whose daily working lives are most directly affected by a digital twin program, yet it is the population least often consulted during program design. The transit authority example that opens this chapter is not exceptional; it is representative of a pattern that plays out in manufacturing, utilities, and logistics programs at organizations of every size. The technical teams that build twins are typically not the operators who use them, and the distance between design and use results in interfaces that answer questions the designers had rather than those the operators actually face.

Effective operator engagement begins before the requirements phase and continues through the full delivery cycle. At the requirements phase, the goal is to understand the actual decision-making process of the operator role: what information they currently use, when they use it, how long they have to act on it, what the consequences of a wrong decision are, and what would make them more confident rather than more confused. In the design phase, prototype interfaces should be tested with operators under realistic workflow conditions, not in conference-room demonstrations. At the delivery phase, operators should be trained by practitioners who can answer questions about why the recommendation is being made, not just how to navigate the interface. Tomás Reyes at the Dayton facility insisted on this sequence. The result was an adoption rate that Renata Vance cited in the quarterly steering committee as the benchmark for subsequent deployments.

Resistance from experienced operators is not irrational and should not be treated as a communication problem to be solved with better messaging. An operator who has spent twenty years developing physical intuition about equipment behavior has genuine expertise that the twin model may not fully capture. The most effective change management approach treats the operator's skepticism as a source of model validation rather than an obstacle to adoption. When operators are given a formal mechanism to flag cases where their judgment diverges from the twins' recommendation, and when those divergences are actually reviewed and used to improve the model, the adoption dynamic shifts from compliance-driven to expertise-collaborative. This is a design decision, not a communications strategy.

Training and Adoption Measurement

Training for digital twin tools must be grounded in workflow context rather than feature coverage. A training curriculum that walks operators through every available interface function in isolation produces tool awareness but not capability. Effective training starts with a realistic operational scenario, introduces the twin interface as it would actually be used during that scenario, and builds from there. Scenario-based training also creates natural assessment opportunities: can the operator correctly interpret the twin's status display under normal operating conditions, under a degraded-mode alert, and under a conflicting signal between sensor data and model prediction? Each of these represents a real decision moment and can be evaluated objectively.

Adoption measurement requires distinguishing access metrics from workflow integration metrics. Login frequency confirms that operators are using the tool; it says nothing about whether the tool is changing decisions. Workflow integration metrics are more meaningful: the proportion of equipment inspections initiated by a twin alert, the frequency with which the twin's pre-shift summary appears in morning handover meetings, and the ratio of operator-flagged model divergences to confirmed errors in review. These metrics require workflow instrumentation, not just tool instrumentation.

Diagram 8.9 - Change Management and Adoption Measurement Framework

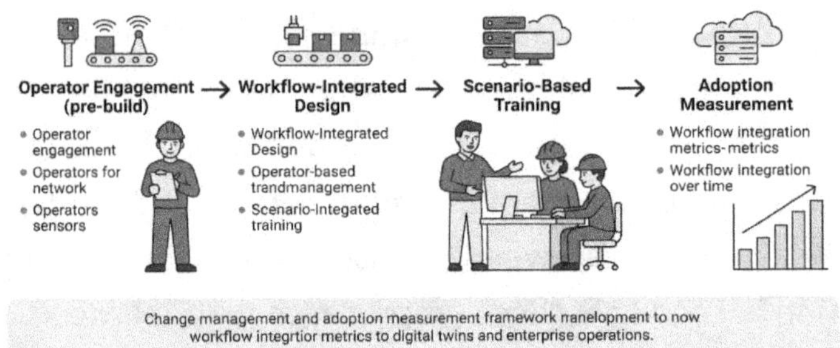

Change management and adoption measurement framework nanelopment to now workflow integrtior metrics to digital twins and enterprise operations.

The leading indicators of program health are organizational rather than financial. They include: operator-initiated improvement suggestions routed to the program team, line supervisor willingness to commit operator time to validation activities, and the frequency with which twin findings appear in operational meeting agendas without prompting. When these behaviors are present, the program has crossed from sponsored adoption to self-sustaining use. When they are absent, the program depends on its original champions and is vulnerable when those champions move on.

NorthArc reached what Renata Vance privately called the 'program escape velocity' threshold at the Dayton facility roughly fourteen months after the first twin deployment. At the monthly operations review in that period, Tomás Reyes opened the session by pulling up the twins' overnight anomaly log rather than the standard maintenance report. No one from the program team had asked him to do it. He had decided it was the most useful thing to put on the screen. That decision, uncoerced and practical, was the clearest evidence that the organizational change had taken hold.

9.8 Manager's Checklist

The following decisions, artifacts, and behaviors define the manager's accountability in the organizational readiness domain of a digital twin program.

- Conduct a formal competency gap analysis before finalizing program scope. Map each of the seven skill domains against current organizational capability and assign a sourcing decision (build, hire, partner, or defer) to each gap.
- Protect domain SME time with a written leadership commitment from the relevant business unit. Do not treat SME availability as a planning assumption; treat it as a negotiated resource allocation.
- Select a team structure archetype (CoE, federated, or hybrid) based on the organization's operational diversity and data function maturity, and document the tradeoffs that informed the choice.
- Designate named liaisons for each operational unit engaged in the program. Fund the liaison role from the program budget. Write a role definition that specifies accountability on both the platform team side and the operational unit side.
- Design a governance committee with four to six named decision-makers, a defined meeting cadence, and explicit authority specifications for platform changes,

use-case approvals, vendor escalations, and safety-related program pauses.

- Evaluate vendor platforms against a scored framework that includes platform breadth, data portability, ecosystem health, services capacity, and lock-in risk. Do not weight demo performance above reference customer evidence.
- Negotiate contract provisions for data rights (what the vendor may use, retain, and share), model ownership (explicit customer ownership of trained artifacts and configurations), and exit provisions (data export formats, transition services, non-interference during disputes).
- Negotiate SLAs that cover data pipeline latency, not just platform uptime. Require notification thresholds and defined remediation timelines for latency breaches.
- Specify knowledge transfer obligations in the integrator contract as a formal deliverable with an acceptance criterion. Do not treat knowledge transfer as an informal courtesy at the close of engagement.
- Build operator engagement into the program plan from the requirements phase. Treat operator skepticism as a model validation input rather than a communications challenge.
- Instrument adoption measurement at the workflow level. Track the proportion of operational decisions informed by twin outputs, not just tool access metrics.

- Define what program escape velocity looks like for the organization: the organizational behaviors that indicate the twin has moved from sponsored adoption to self-sustaining use. Make this definition explicit before deployment, so you can recognize it when it happens.

9.9 Takeaway

A digital twin program can have a sound technical architecture, a well-chosen platform, and a capable data engineering team and still fail if the organizational conditions around it are not deliberately designed. The organizational conditions that determine program outcomes are: a competency base that covers all seven skill domains without critical gaps, a team structure that keeps delivery capacity close to operations, liaison roles that translate requirements without distortion, vendors selected on long-term partnership criteria rather than demonstration performance, contracts that protect data rights and model ownership, integrators held to explicit knowledge transfer obligations, and a change management approach that treats operators as collaborators rather than users.

Priscilla Okonkwo and Hector Salinas left the last vendor session with a recommendation that surprised the program team: they did not select the vendor with the most impressive demonstration. They selected the vendor whose reference customers described the most responsive escalation process, whose contract team had offered the cleanest data portability provisions without being asked, and whose proposed architecture team had direct experience with NorthArc's specific historian platform. The decision was made on evidence of partnership durability, not on the quality of the rendered visualization. That discipline, applied consistently to every organizational decision in the readiness domain, is what separates twin programs that endure from those that do not.

10 From One Twin to an Enterprise Program: Scaling, Federation, and the Road Ahead

10.1 The Boardroom Moment

Renata Vance had prepared for the board presentation the way she prepared for every high-stakes moment: with one slide she could defend for twenty minutes if the directors asked her to stop there. The slide showed a three-year roadmap. Year one was a column of four operational twins already live across NorthArc Manufacturing and Energy Services. Year two showed those twins linked through a shared semantic layer. Year three showed a federated enterprise twin network spanning all three operating units, plugged into the sustainability reporting engine, and informing real-time capital allocation decisions. The board asked her to stop at that slide.

After the meeting, the NorthArc team gathered in the conference room adjoining the boardroom. Hector Salinas noted that the Dayton pilot had nearly died twice: first when the plant historian could not feed data fast enough to keep the twin synchronized, and again when the production team concluded the twin's anomaly alerts were generating more noise than signal. Tomás Reyes, joining by phone, confirmed both moments. Dorian Whitlow said that was the pattern at every enterprise he had studied. The pilot-to-production gap was not a technology gap; it was a trust gap, a workflow gap, and a governance gap arriving simultaneously. What Renata now had to articulate was not what digital twins were, but how NorthArc had learned to make them stick.

10.2The Pilot-to-Production Attrition Problem

Six Failure Modes

Most digital twin pilots that demonstrate positive results do not reach full production deployment. The failure modes are consistent across industries and are addressable if named early.

Integration debt is the first mode. Pilots built with direct point-to-point connections require full custom integration work at each new site. The cost accumulates faster than the business value, and the program stalls. Tomás Reyes saw this at Dayton: the historian connection that took three weeks for one production line would have taken three weeks per line without a reusable connector framework.

Trust collapse follows a high-profile false positive. When the twin flags a predicted failure that does not materialize, operators update their mental model. Dorian Whitlow's team tracked the recommendation acceptance rate: the fraction of twin-generated recommendations that operators acted on. When that rate dropped below forty percent, it was a leading indicator of impending abandonment, regardless of model accuracy metrics.

Governance vacuum is the third mode. A pilot runs under the personal sponsorship of one executive. When that person changes roles or loses budget authority, the pilot loses its organizational protector. Without a formal steering committee, defined budget lines, and written decision rights, there is no structure to sustain the program across leadership transitions.

Platform lock-in without strategic alignment creates the fourth failure. Organizations that choose a twin platform based on a single use case discover that the platform's data model or licensing structure creates friction for adjacent use cases. Priscilla Okonkwo required every platform shortlist to answer two questions: what does expansion to ten times the current sensor volume cost, and what is the documented migration path if NorthArc switches platforms in year four.

Data quality debt hidden during the pilot is the fifth mode. Well-managed pilot sites often have cleaner data than the enterprise average. Dr. Mei Lin Park called this the Potemkin village problem: the pilot looked representative, but it was not. Remediation plans need to be part of the scaling architecture, not afterthoughts.

The sixth mode is the absence of a reusable operating model. A pilot is run by a dedicated team that knows every quirk of the deployment. When the program scales, that team cannot replicate itself. If runbooks, alert logic, escalation paths, and model refresh procedures are not documented, the next deployment starts from scratch. Tomás Reyes spent four months after the Dayton pilot writing the playbook that became the template for every subsequent NorthArc twin deployment.

Diagram 9.1 - Pilot-to-Production Attrition: Six Failure Modes and Intervention Points

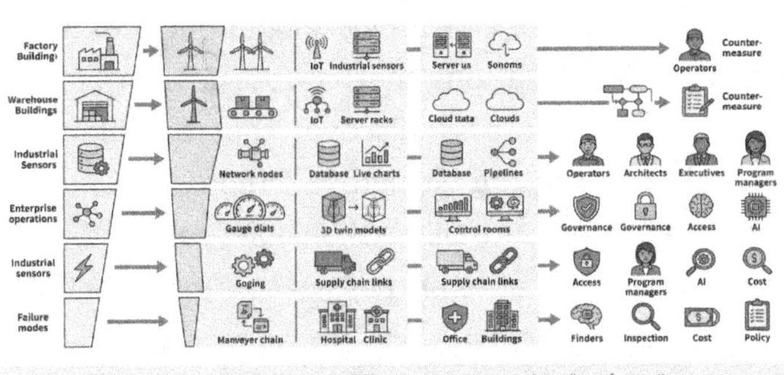

Pilot-to-Production attrition: Six failure modes and intervention points strong a print-quality arc for operations.

Two Near-Deaths Revisited

The first near-death at Dayton was a technology problem with a technology solution. The PI System historian had not been configured to support the polling frequency required by the twin. The initial integration pulled data at one-minute intervals; the anomaly detection models needed a ten-second resolution to catch the vibration signatures preceding motor failures. A historian reconfiguration and network upgrade, taking six weeks, resolved the issue. The lesson: data infrastructure readiness needs assessment and remediation before twin platform deployment, not during.

The second near-death was entirely human. By month four, the twin was generating an average of 11 anomaly alerts per shift. Operators investigated the first dozen; seven yielded no actionable issues, so they began routing alerts to a low-priority queue. Tomás Reyes raised the issue; Dorian Whitlow's team discovered the alert threshold had been set for model showcase, not operator workflow. Three weeks of recalibration reduced alerts to two per shift, with a confirmation rate above 80%. Within sixty days, operator trust scores recovered to pilot levels. The lesson: the right technical accuracy is the accuracy level that sustains human engagement, and those two numbers are not the same.

10.3Patterns That Scale: The Architectural Foundation

Reusable Data Contracts

A data contract is a formal, machine-readable specification of what a source system agrees to provide: field names, types, units, refresh frequency, quality standards, and conditions under which the contract can be modified. In a single-twin pilot, contracts are usually implicit. At a dozen twins across three operating units, implicit contracts collapse.

Dr. Mei Lin Park's architecture team spent the first quarter after Dayton building a data contract registry. Every source system a twin would consume had to register a contract before integration work began. The contract specified schema and SLA: minimum uptime, maximum latency, and notification window for schema changes. When a historian at the Cincinnati facility changed the tag-naming convention during a vendor upgrade, the registry automatically flagged the change. It generated an impact assessment showing which twin models would be affected. Without the registry, that change would have surfaced as unexplained model drift weeks later.

Reusable contracts also significantly reduce the effort required for new-site integration. NorthArc's team estimated a roughly 60% reduction compared to bespoke integration because each source system type needed documentation only once. A new manufacturing plant could inherit Dayton's contract templates, override the site-specific parameters, and begin twin deployment with a pre-validated integration foundation.

Diagram 9.2 - Reusable Data Contract Registry: Structure, Lifecycle, and Scaling Benefit

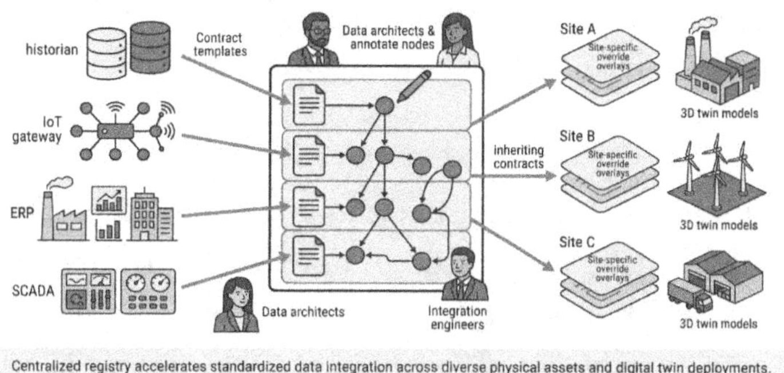

Centralized registry accelerates standardized data integration across diverse physical assets and digital twin deployments.

The Shared Semantic Model

A shared semantic model defines what assets, processes, relationships, and measurements mean across the enterprise. Without one, cross-unit analytics become translation projects. When Renata Vance wanted a single enterprise dashboard showing maintenance cost per unit of throughput across all three operating units, the semantic divergence between Manufacturing's asset-type classification and Energy Services's work-order-category classification became immediately visible. The resolution required a mapping layer in the semantic model, not a forced unification of operational taxonomies. Semantic models must be built through negotiation across business units, with an explicit representation of when mapping is the right answer and when forced standardization is wrong.

NorthArc's architecture team adopted the Asset Administration Shell standard as the semantic foundation for manufacturing assets, while building custom extensions for energy services and logistics assets that lack standard coverage. Dr. Mei Lin Park's rule: use a standard where one exists and is fit for purpose; extend it where it is not; invent only when there is no alternative.

Platform of Platforms

The requirements for asset twins in a manufacturing plant differ substantially from those of network twins modeling a power distribution grid, which differ from those of process twins modeling a logistics network. Forcing all use cases onto a single platform creates one of two outcomes: either the platform is comprehensive enough to serve all use cases but prohibitively complex and expensive for simpler ones, or it is optimized for the dominant use case and underserves the others.

NorthArc's architecture resolved this with a platform-of-platforms model. Each operating unit retained the right to select the platform best suited to its primary use cases, subject to three non-negotiable constraints: data contract compliance, participation in the shared semantic layer through a standard API, and compliance with enterprise security and audit requirements. NorthArc Manufacturing used Siemens Xcelerator for plant asset twins. Energy Services combined Azure Digital Twins for network modeling with GE Digital for turbine-level twins. Logistics used AWS IoT TwinMaker for the distribution-center process twins. All three platforms fed into the same enterprise semantic layer.

Priscilla Okonkwo's team built the vendor governance framework that made this manageable. Each vendor had quarterly business reviews covering integration health, roadmap alignment, and SLA performance under contract. All contracts included data portability provisions: NorthArc retained ownership of all twin models, data, and configurations, and vendors were obligated to provide migration assistance if NorthArc chose to change platforms.

Diagram 9.3 - Platform of Platforms: Multi-Vendor Twin Architecture with Shared Semantic Layer

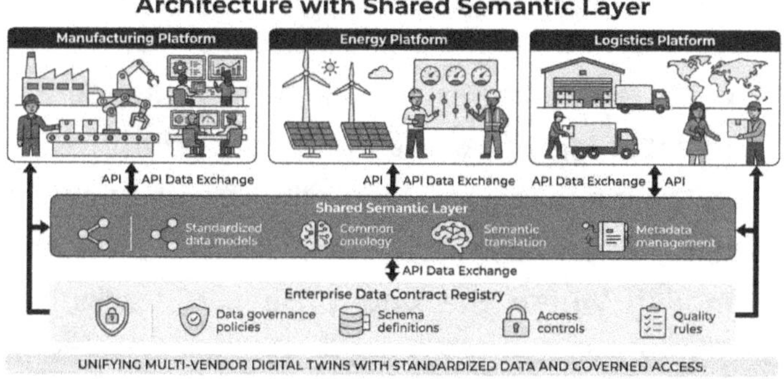

10.4Federation Across Business Units and Geographies

What Federation Means in Practice

Federation is an architectural and governance model in which individual twins retain operational autonomy while contributing to a shared enterprise capability. A federated program is not a centralized program with distributed execution points; it is a network of semi-autonomous programs that agree on interfaces, semantics, and governance standards while retaining local implementation freedom within those constraints.

The first eighteen months of NorthArc's program operated under a quasi-centralized model, with Dr. Mei Lin Park's team controlling all platform and integration decisions. When Hector Salinas's manufacturing team needed a plant simulation capability that the central platform did not support, the architecture review process took four months. By the time approval came through, two members of his operational improvement team had left in frustration. The shift to a federated model moved the architecture team's role from gatekeeper to enabler: publishing pattern libraries, running the data contract registry, maintaining the semantic model, and providing integration tooling that units adopted rather than custom-built. Renata Vance described this as moving from a railroad model, where the enterprise lays tracks, to a highway model, where the enterprise builds roads and safety standards but allows different vehicles to travel at their own pace.

Diagram 9.4 - Federation Model: Centralized vs. Federated Twin Governance Architecture

Federation Model: Centralized vs. Federated Twin Governance Architecture

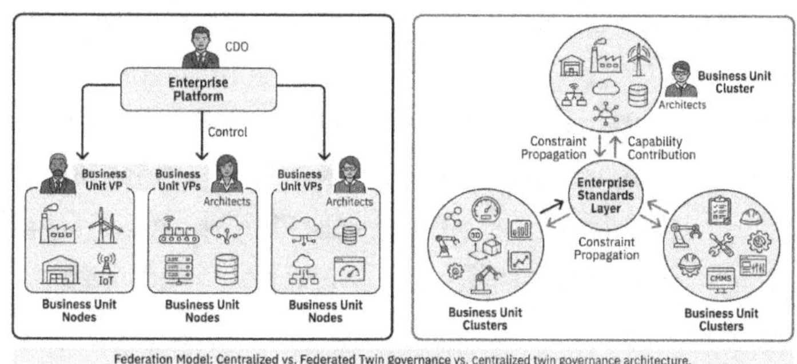

Federation Model: Centralized vs. Federated Twin governance vs. centralized twin governance architecture.

Geographic Federation and Cross-Border Data

For enterprises operating across multiple countries, federation introduces regulatory complexity. A digital twin aggregating operational data from assets in different jurisdictions operates under distinct regimes governing data localization, privacy, and in some sectors, critical infrastructure protection. Some regulatory regimes treat AI-generated recommendations based on operational data as subject to the same residency requirements as the underlying data.

NorthArc Energy Services had a joint venture in Ontario, Canada, creating cross-border data flows. Aisha Bramwell's team mapped the regulatory requirements before any data contracts were finalized. The outcome was a federated data architecture where Ontario operational data was processed and modeled locally, with only anonymized performance metrics flowing to the enterprise layer. The governance lesson: cross-border data flows need to be mapped in the enterprise data contract registry as a first-class attribute, not an annotation. Every NorthArc contract included a jurisdiction field. The enterprise analytics layer had a policy engine that checked jurisdiction constraints before running cross-system queries or simulations, either blocking non-compliant queries or applying the anonymization required for compliance.

10.5The Twin-of-Twins: Enterprise-Level Situational Awareness

The Concept and Value Proposition

A twin-of-twins is an enterprise model that aggregates the outputs of individual asset, process, and system twins to provide cross-portfolio visibility, cross-domain simulation, and enterprise-level decision support. The first question Renata Vance wanted to answer: if the Dayton plant experiences an unplanned two-day outage, what is the optimal reallocation of production and logistics capacity across the enterprise to minimize customer impact and cost?

That question required data from Manufacturing's plant twins (inventory and production capacity at each facility), Logistics's process twins (routing and capacity), and Energy Services's demand-forecasting models (energy-cost implications of operating alternative facilities at higher utilization). No single operational team could answer it manually in under a day. The twin-of-twins, once integrated with all three sources, could run the optimization simulation in under thirty minutes and present results in a decision-ready format. The first time Dorian Whitlow demonstrated this capability, Hector Salinas said it was the first time in twenty-two years of manufacturing that he had seen a system tell operations leaders what they needed to know before the situation had already resolved itself.

Diagram 9.5 - Twin-of-Twins Architecture: Individual Twins Feeding Enterprise Simulation Layer

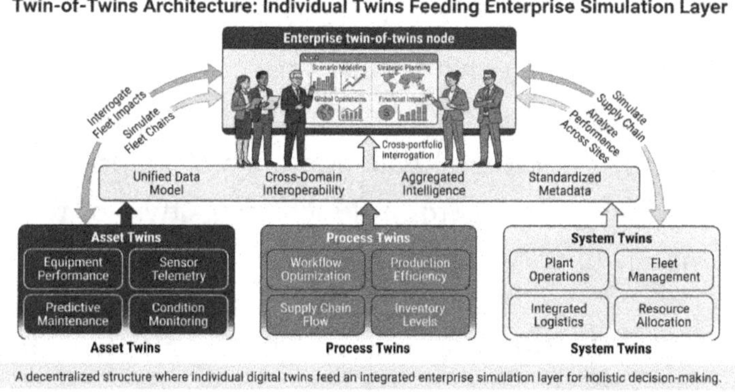

A decentralized structure where individual digital twins feed an integrated enterprise simulation layer for holistic decision-making.

Sequencing Prerequisites

The twin-of-twins is a destination, not a starting point. It requires individual twins at operational maturity and a validated shared semantic model before it can deliver reliable value. Attempting enterprise aggregation before those foundations are stable produces a system that is technically impressive and operationally useless. Dr. Mei Lin Park's principle: a twin-of-twins built on shaky foundations inherits every data quality problem, every integration gap, and every model accuracy issue from its constituent twins. The enterprise layer amplifies those problems; it does not average them out.

NorthArc's sequencing followed three stages. Stage one established individual twins at operational maturity, defined as a consistent recommendation acceptance rate above 60% and a verified data synchronization SLA of 99% uptime. Stage two built the shared semantic model and data contract registry, and validated it through cross-unit benchmarking queries, which all three business unit heads confirmed produced accurate results. Stage three implemented the twin-of-twins on those validated foundations, starting with read-only aggregation and progressing to simulation and optimization capability after six months of operational validation. Total elapsed time from the Dayton pilot to a functioning twin-of-twins: approximately twenty-eight months.

10.6Sequencing the Enterprise Roadmap: Year One Through Year Three

Year One: Foundations and Proof of Replication

Year one should accomplish three things: prove that the pilot use case generates real operational value, demonstrate that the architecture can support a second deployment without full custom rebuilding, and establish governance structures that will sustain the program beyond its founding sponsorship. These are minimum conditions for year-two investment to be rationally approved, not aspirational targets.

NorthArc's year-one measurement framework tracked four metrics per twin: mean time between unplanned failures compared to the twelve-month pre-twin baseline, maintenance cost per unit of throughput, recommendation acceptance rate, and data synchronization SLA compliance. The proof of replication is the most commonly underweighted year-one objective. NorthArc's year-one plan included a second deployment at a Cincinnati facility, explicitly designed as a replication test. Using Dayton's contract templates and operating model, the second deployment went live three months after Dayton and identified four integration patterns that needed generalization before reliable multi-site use.

Diagram 9.6 - Enterprise Twin Roadmap: Year One Through Year Three Milestone Architecture

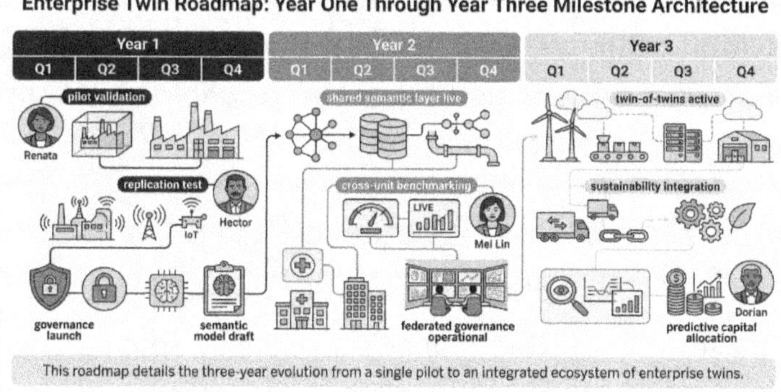

Year Two: Integration and Federation

Year two is the most organizationally demanding phase. The program moves from a project with a defined scope and a clear finish line to a program with ongoing operations, multiple active deployments, and governance that must balance competing demands across different business units. The organizational demands are larger than the technical demands, and most programs that stall do so here for organizational rather than technical reasons.

Year-two data quality remediation typically reveals its full cost at this stage. The Potemkin village effect arises as the program expands to sites not selected for the pilot due to data quality challenges. NorthArc's year-two data quality investment was roughly 40% larger than budgeted, primarily because three manufacturing facilities had historian configurations that had not been updated in six or more years. Dr. Mei Lin Park's team built a data quality assessment tool deployed to every facility before any twin onboarding began, producing a readiness score and remediation estimate. Facilities below a defined threshold were put on a remediation track before deployment, not during.

Year Three: Enterprise Capability and Strategic Integration

Year three marks the transition from a digital twin program to a digital twin capability. A program has a team, a budget, and deliverables. A capability is embedded in the operational fabric: the twin is the system of record for asset health, the simulation layer is the standard tool for operational planning, and the enterprise aggregation layer is the foundation for capital allocation analysis. The distinction determines whether the program survives a leadership transition or a budget reduction.

Aisha Bramwell recognized that the operational data NorthArc already collected for twin purposes was the same data required for accurate GHG reporting, and that most enterprises paid significant consulting fees to collect it manually through separate processes. Connecting the twin data layer to the sustainability reporting engine eliminated manual collection and improved accuracy. When the CFO's team could run a capital allocation scenario through the twin-of-twins simulation layer and see projected operational impact before committing capital, the program moved from a technology initiative to a strategic planning tool. That transition is what earns a twin program permanent budget status.

10.7 Portfolio Governance: Managing Many Twins at Once

The Twin Inventory and Program Registry

When a program reaches 10 or more active deployments, it requires a management layer. With twenty active twins across three business units, the governance questions multiply rapidly. Which twins are operating within their data synchronization SLAs? Which model versions are running, and which have drifted from the master registry?

Which deployments have fallen below the recommendation acceptance threshold that indicates operator disengagement? Without a program-level registry that continuously answers these questions, program leadership is managing by anecdote.

NorthArc's twin program registry tracked four health dimensions for each active twin: operational health (synchronization SLA, data quality scores, alert volumes), model health (version currency, drift metrics, last validation date), adoption health (recommendation acceptance rate, active user count, support ticket volume), and governance health (data contract compliance, security certification currency, audit finding status). Dr. Mei Lin Park's team reviewed the registry weekly and escalated items below defined thresholds to the steering committee's monthly report. The registry also served as the intake mechanism for new twin requests, requiring use case, anticipated data sources, estimated data readiness, proposed platform, and business value metrics before any onboarding work began.

Diagram 9.7 - Portfolio Governance Dashboard: Twin Inventory Registry with Four Health Dimensions

Portfolio Governance Dashboard: Twin Inventory Registry with Four Health Dimensions

Model Lifecycle Governance at Portfolio Scale

Model lifecycle governance manages the creation, validation, deployment, monitoring, refresh, and retirement of AI and simulation models across the portfolio. At portfolio scale, the challenge grows nonlinearly because models are shared across deployments, refreshed at different times, and subject to drift that may affect some deployments but not others.

Dorian Whitlow's team built the framework around three principles: a canonical model version in the enterprise registry (deployments ran named versions, upgrades required a formal change process with rollback capability); per-deployment drift monitoring (drift beyond a threshold triggered a model review flag); and scheduled model retirement (every model had a maximum operational lifetime requiring full revalidation). The version governance system earned its cost in year two when a physics model update intended for manufacturing turbine twins was inadvertently propagated to energy services turbine twins. The configurations differed in ways that made the update inappropriate for energy services. The governance system flagged the cross-unit propagation before it completed, preventing a model mismatch that would have affected energy dispatch decisions.

10.8What Comes Next: The Technology Horizon

AI Agents Operating Twins

The current generation of digital twins supports human decision-making. The next generation will feature AI agents that not only support decisions but also execute them autonomously within defined operational boundaries. An AI agent continuously monitors the twins' state, identifies decision opportunities, evaluates response options through simulation, selects the highest-value option, and executes it via direct integration with the control layer, all without human intervention within its authority boundary.

The governance framework for AI agents is more demanding than for AI-assisted twins. When an agent executes autonomously, the accountability chain must be documented in advance: which decisions the agent is authorized to make, under what conditions human escalation is required, how the decision log is retained for audit, and what the rollback procedure is for unintended outcomes. Aisha Bramwell's team began drafting

NorthArc's AI agent governance framework in year three, drawing on NIST AI RMF guidance and the emerging ISO/IEC 42001 AI management system standard. Managers skeptical of autonomy should note that the authority boundary is fully configurable: the practical benefit of bounded AI agent autonomy is capturing the value of rapid response to conditions that occur faster than human reaction cycles allow, while preserving human authority over decisions requiring organizational judgment.

Diagram 9.8 - AI Agent Autonomy Spectrum: From Advisory Twin to Autonomous Operational Agent

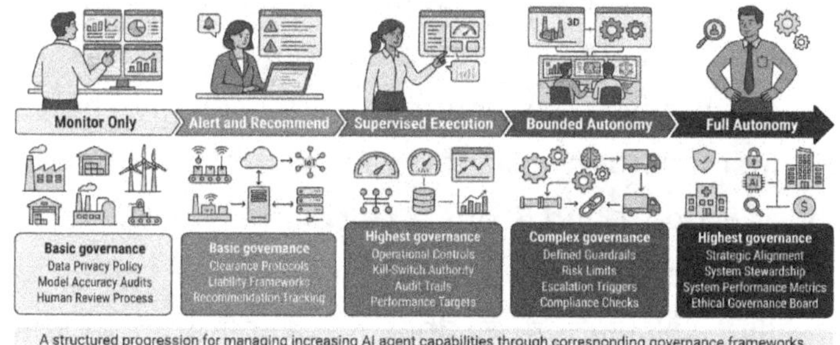

A structured progression for managing increasing AI agent capabilities through corresponding governance frameworks.

Generative Simulation, Sustainability Twins, and Urban Twins

Generative simulation uses large generative AI models to create novel operational scenarios not observed in historical data. Traditional simulation requires an engineer to define scenario parameters manually. Generative simulation inverts this: the AI generates a library of plausible scenarios based on physical models, operational history, and risk frameworks, presenting the most consequential ones for simulation. Dorian Whitlow's team was developing a scenario library to simulate supply chain disruptions, extreme weather events, and simultaneous equipment failures across multiple facilities. Renata Vance's view was that generative simulation would transform the way NorthArc conducted its annual risk assessment, systematically exploring a scenario space that historical data and expert judgment alone could never cover.

Sustainability twins use the data infrastructure of operational digital twins to generate accurate, audit-ready emissions and resource consumption data for ESG reporting. NorthArc's sustainability twin initiative, launched in year three, demonstrated the integration path: the energy consumption data already flowing through manufacturing plant twins was the same data needed for Scope 1 and Scope 2 calculations at the facility level. By connecting the existing twin data layer to a sustainability calculation engine and audit trail module, NorthArc replaced three months of manual data collection with a continuous automated process. Aisha Bramwell's team validated the automated reporting against the GHG Protocol standard with NorthArc's external sustainability auditor. The sustainability twin also opened a new management use case: operational decisions with emissions optimization as a secondary objective. Production reallocation scenarios that appeared cost-optimal on a pure production basis were sometimes suboptimal when carbon cost was included, and vice versa.

The scope of digital twin thinking has also expanded beyond individual enterprises. Cities, ports, airports, and utility networks are building twins that model complex interdependencies at scales beyond the control of any single organization. For enterprise leaders, the relevant question is how the enterprise's own twin infrastructure will interface with public infrastructure twins. If a port twin or freight network twin exposes a data API, NorthArc's logistics twin could consume disruption forecasts and route optimization recommendations from the external system without building or maintaining the port-level model itself. The data contract framework built for internal integration is exactly the framework needed to consume external infrastructure twin data. Dr. Mei Lin Park had already drafted a template for external twin data contracts to support this use case.

10.9How Managers Stay Current in a Moving Field

The digital twin field is evolving faster than any single program can track comprehensively. Platform capabilities expand quarterly. New AI integration patterns emerge from research and from industry deployments. Regulatory frameworks governing AI autonomy and data governance are being revised in real time. A manager responsible for an enterprise digital twin program cannot track all of this through passive awareness; they need a structured approach that filters signals from noise and connects emerging developments to specific program decisions.

The most effective practice the NorthArc team developed was the quarterly technology review. Every quarter, Dorian Whitlow's team prepared a one-page summary of significant developments in four areas: platform capability releases from active vendors, research publications from leading digital twin research groups, peer case studies from manufacturing and energy sector enterprises, and regulatory updates from monitored standards bodies (ISO TC184, IIC, NIST). The summary was a curated assessment of developments most relevant to NorthArc's roadmap decisions in the next twelve months, not a comprehensive survey. It was presented to the steering committee and generated an action register of items requiring architectural or governance response.

Professional communities of practice are a second valuable channel. The Digital Twin Consortium and the Industrial Internet Consortium publish working group outputs, reference architectures, and peer case studies unavailable through commercial research channels. Renata Vance's team maintained memberships in three such bodies. It rotated technical staff through working groups, gaining peer intelligence and the ability to shape emerging standards toward patterns NorthArc could implement without major redesign. Vendor roadmap briefings are a third channel, one requiring calibration. Priscilla Okonkwo treated vendor briefings as one input among many, validated against peer enterprise experience and analyst research before any architectural or procurement decision.

Diagram 9.9 - Technology Horizon Scanning Framework: Four Input Channels and Quarterly Decision Cadence

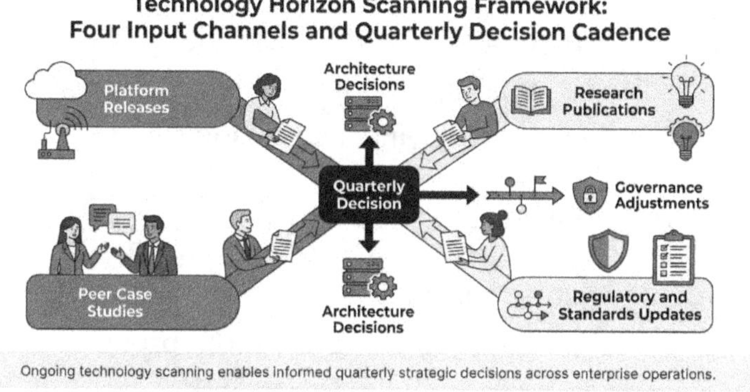

10.10 Manager's Checklist for Scaling to an Enterprise Program

The decisions, artifacts, and behaviors below are the minimum governance posture that distinguishes programs that sustain and scale from programs that plateau and fade.

Pilot-to-Production Foundations:

- Define pilot success metrics before deployment, including recommendation acceptance rate as a primary metric alongside technical accuracy measures.

- Conduct a replication test with a second deployment before scaling beyond two sites. The replication test validates whether architectural decisions generalize.
- Document the pilot operating model: runbooks, alert thresholds, escalation paths, model refresh procedures. This is the program's most transferable asset.
- Assess data readiness at every candidate site with a standardized scoring tool before onboarding. Remediate before deploying, not during.

Architecture Artifacts:

- Maintain a data contract registry with jurisdiction, schema, SLA, and change notification provisions for every source system in the twin network.
- Build and maintain a shared semantic model covering all asset types in the portfolio. Document mapping layers explicitly where forced unification is the wrong choice.
- In a platform-of-platforms model, enforce three non-negotiables for all unit platforms: data contract compliance, semantic layer participation via standard API, and enterprise security certification.
- Include data portability provisions in all platform vendor contracts. Retain ownership of all models, configurations, and data.

Federation and Governance:

- Launch a steering committee with defined decision rights before the program scales beyond three active deployments. Include business unit representation from the start.
- Build a program-level twin registry tracking operational, model, adoption, and governance health for every active deployment.
- Implement model version governance with a canonical registry, version-specific deployment tracking, and automated drift monitoring.
- Map cross-border data flows as a first-class registry attribute. Implement a policy engine to enforce jurisdictional constraints before executing cross-system queries.

Roadmap and Portfolio:

- Publish a three-year roadmap with explicit milestones for foundation, integration, and enterprise capability phases. Tie year-over-year budget approvals to milestone outcomes.
- Sequence the twin-of-twins only after individual twins reach operational maturity and the semantic model is validated by cross-unit analytics consumers.
- Run a quarterly technology horizon scan covering platform releases, research, peer case studies, and regulatory updates. Maintain an action register for items requiring response.
- Integrate sustainability data flows into the twin architecture by year two or three. The operational data

already collected overlaps substantially with GHG reporting requirements.

- Draft AI agent governance boundaries before the first autonomous execution capability is deployed. The governance framework takes longer to build than the technology.

10.11 Takeaway

The gap between a successful digital twin pilot and a sustainable enterprise program is real, predictable, and survivable. It is not a technology gap; it is an architectural discipline gap, a governance maturity gap, and an organizational trust gap, and all three are addressable with deliberate design choices made early. The managers and organizations that navigate this gap successfully share a common pattern: they treat the second deployment as seriously as the first; they build shared infrastructure before they need it; they establish governance before the conflicts that require it arise; and they measure adoption outcomes, not just technical metrics.

The NorthArc program Renata Vance presented to the board was not the product of exceptional technology or exceptional luck. It was the product of a team willing to name the failure modes, learn from near-death experiences, build the architecture that would not fail the same way twice, and govern the program with enough rigor to protect it across budget cycles and leadership transitions. The technology will continue to evolve, the use case space will expand, and the governance frameworks will be refined. What will not change is the fundamental requirement that enterprise digital twin programs be built on operational trust, shared infrastructure, and governance discipline. Those foundations are available to any organization willing to invest in them before they are needed, rather than discovering they are absent when they matter most.

11 Closing

Three years ago, Renata Vance stood before NorthArc's board and proposed spending eighteen months and considerable capital on a digital twin program that most of the room considered an expensive experiment. The turbine failure at Columbus had just cost eleven days of lost production and a regulatory notice. Renata's argument was direct: the real cost was not the downtime but the ignorance that allowed it. No one had seen the failure coming because no one had a reliable, continuously updated picture of that asset's operating state.

Today, that reluctance is difficult to find. NorthArc's program has grown from one turbine asset twin to a coordinated portfolio spanning five manufacturing facilities, two energy services substations, and the Dayton logistics hub. Hector Salinas now runs onboarding workshops for plant managers across the enterprise; when the predictive model caught a bearing anomaly on Line 4 twelve days before failure, the avoided downtime exceeded the full cost of the Dayton pilot. Mei Lin Park's data fabric connects historical data, ERP records, and IoT telemetry across every active twin, with integrations now taking three weeks rather than three months. Dorian Whitlow's model inventory tracks thirty-one active predictive models, each with a refresh cycle and a named owner, a standard Aisha Bramwell required before any model could be called decision-critical. Tomás Reyes contributed the runbook redesigns that made twin outputs usable on the floor, and Priscilla Okonkwo's vendor frameworks ensure no single platform holds an exit-blocking position. What these seven leaders built is more durable than any individual deployment: a governance capability that will outlast every platform version it currently runs on.

11.1The Themes That Ran Through Every Chapter

Nine chapters covered the territory from definitional clarity to change management. Beneath the specifics, four principles surfaced repeatedly because they define programs, not chapters. Use-case primacy: every worthwhile twin decision starts with a specific operational problem carrying a named owner and a measurable current cost; programs that begin with platform selection before that definition is complete consistently underperform. Data governance as infrastructure: quality is an organizational responsibility, not a technical task deferred to engineers; every program that has scaled past pilot phase has done so because someone with authority owned the data architecture and had the mandate to enforce it. Value in decisions, not dashboards: a twin that does not change a maintenance call, a scheduling run, or a resource allocation decision is an expensive monitoring screen, and outcome measurement is the ongoing mechanism by which the program justifies continued investment. Responsible-by-design: governance, auditability, and model oversight are architectural requirements from day one, not compliance layers applied after the fact.

Diagram C.1 - The Manager's Transformation Arc

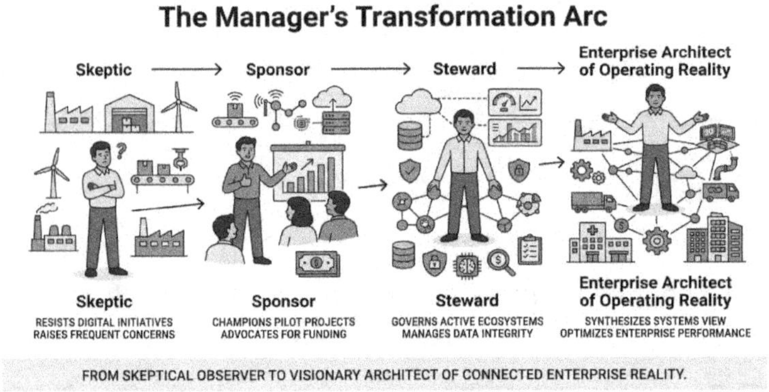

Most managers who read this book entered their twin journey in the skeptic phase: unconvinced that the technology was mature, uncertain whether the ROI case was real, wary of committing organizational credibility to a program with uncertain outcomes. Skepticism at that stage is sound governance. The appropriate response to an unproven capability is rigorous scoping, disciplined pilot design, and clear outcome criteria before any production investment is made.

The transition from skeptic to sponsor happens when pilot evidence is credible, and the use-case owner has a documented outcome story. Sponsorship is organizational commitment backed by budget authority, not enthusiasm. Sponsors who advance a program without also building stewardship capacity tend to see it plateau, because governance requirements at production scale differ qualitatively from those at pilot scale. Stewardship means building the data ownership structures, model oversight calendars, and change management disciplines that keep a production twin program mission-aligned over time. The enterprise architect of operating reality is a managerial posture: the leader who governs both the physical and digital representations of operations as a unified, critical infrastructure responsibility.

11.2The Road Ahead

Three structural forces will reshape enterprise twin programs over the next three years. Generative AI integration is moving from demonstration to production as query interfaces and simulation co-pilots become standard platform features; managers with strong model governance frameworks will absorb these capabilities without creating audit exposure because the oversight structures for predictive models and generative tools are structurally similar. Twin interoperability will shift from aspiration to expectation as Asset Administration Shell adoption accelerates and customers issue requirements that single-vendor architectures cannot meet. The regulatory posture around operational AI will tighten across energy, financial services, healthcare, and aviation sectors as authorities move toward documented governance requirements for decision-critical AI outputs. Organizations that built accountability in from the start will find this transition manageable; those that deferred it will face significant remediation costs.

Diagram C.2 - Three-Year Horizon: Where Enterprise Twin Programs Are Going

Three-Year Horizon: Where Enterprise Twin Programs Are Going

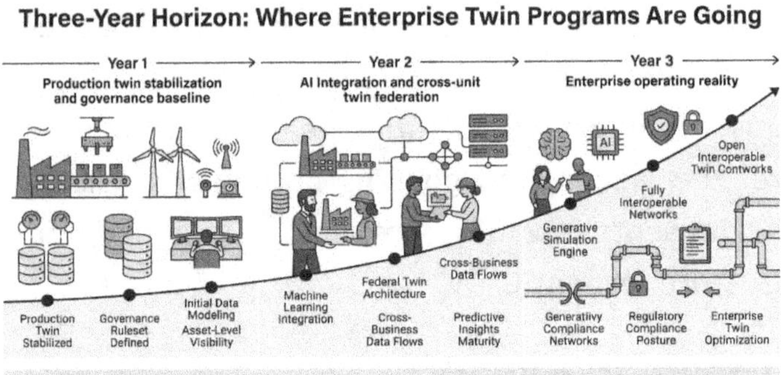

A roadmap for advancing enterprise twin programs from production stability to integrated, generative operating reality.

11.3A Practical Call to Action

This book has been organized around decisions rather than descriptions because the manager's role is to create the organizational conditions under which the technology delivers its intended value: clear use-case ownership, funded and maintained data governance, AI oversight that satisfies both internal risk management and external regulatory requirements, vendor relationships that preserve strategic flexibility, and adoption disciplines that connect twin outputs to the workflows of the people who depend on them.

If you are in the early stages of a program, slow down the platform conversation and invest time in use-case definition, data quality assessment, and outcome measurement design before any architecture decision is final. Organizations that succeeded at scale did so not because they chose the right platform but because they defined the right problem, governed the data on which the solution depended, and measured outcomes rigorously enough to make the ROI case compelling to the stakeholders controlling the next phase of investment.

If you are managing a program already in production, run a governance audit: review data ownership for gaps, confirm the model inventory is current with a documented performance baseline and named reviewer for every decision-critical model, assess vendor contracts for exit-blocking dependencies, and verify that adoption metrics capture workflow-level impact rather than platform utilization statistics.

Renata's original bet was on management discipline over technology. The turbine that failed in 2024 will eventually be retired. The governance capabilities, the data fabric, the model oversight structures, and the operational trust that NorthArc built around its twin program will outlast any individual asset or platform version. Those capabilities are the real deliverables of a well-governed twin program. They are what it means to operate the future first.

12 References

Alam, Khairul, et al. "Digital Twin–Driven Smart Manufacturing: Recent Advances and Future Trends." *Journal of Manufacturing Systems*, 2024. (CH3)

Allam, Zaheer, and David Jones. "Digital Twins for Cities: From Infrastructure Insight to Urban Operating Systems." *Cities*, 2023. (CH4)

Bosch Rexroth. *Digital Twin in Manufacturing: From Asset Models to Closed-Loop Optimization*. Bosch Rexroth AG, 2024. (CH2)

Broo, Daniel, et al. "Governance of Digital Twins in Safety-Critical Systems." *Safety Science*, 2023. (CH8)

Cai, Yujia, et al. "A Reference Architecture for Industrial Digital Twins in Process Industries." *Computers in Industry*, 2024. (CH3)

Deloitte. *Scaling Digital Twins: From Pilot Factories to Enterprise Platforms*. Deloitte Insights, 2023. (CH10)

European Union Agency for Cybersecurity. *Cybersecurity for Digital Twins in Critical Infrastructure*. ENISA, 2024. (CH8)

European Commission. *Digital Twins and Data Spaces: Building the Foundations for Industrial Metaverse*. Publications Office of the European Union, 2025. (CH6)

Gartner. *Hype Cycle for Digital Twins and Industrial Metaverse, 2024*. Gartner Research, 2024. (CH1)

IBM Institute for Business Value. *Digital Twins and AI: Turning Operational Data into Continuous Decision Systems*. IBM, 2023. (CH7)

International Organization for Standardization. *ISO 23247: Digital Twin Framework for Manufacturing*. ISO, 2023. (CH2)

International Organization for Standardization. *ISO/IEC 42001: Artificial Intelligence Management System—Requirements*. ISO, 2024. (CH8)

Kritzinger, Werner, et al. "Industrial Digital Twin Architectures: Interoperability, Integration, and Lifecycle Management." *IEEE Access*, 2023. (CH6)

Lee, Jay, et al. "Industrial AI and Digital Twins: Architectures for Predictive and Prescriptive Operations." *Procedia CIRP*, 2023. (CH7)

Li, Xiaoming, et al. "Data Governance Frameworks for Digital Twin–Enabled Smart Factories." *Robotics and Computer-Integrated Manufacturing*, 2024. (CH8)

Liu, Yifan, et al. "Digital Twin–Enabled Predictive Maintenance in Power Generation Assets." *Energy Reports*, 2023. (CH4)

McKinsey & Company. *Capturing Value from Digital Twins: Business Cases, ROI, and Scaling Patterns*. McKinsey Analytics, 2023. (CH5)

Microsoft. *From Real-Time Data to Digital Twins: Azure Patterns for Industrial Programs*. Microsoft, 2023. (CH3)

Noh, Sang Do, and Seyed Mohammad Mehdi Sajadieh. "From Simulation to Autonomy: Integration of Artificial

Intelligence and Digital Twins." *International Journal of Precision Engineering and Manufacturing–Green Technology*, 2025. (CH7)

PricewaterhouseCoopers. *Digital Twin Governance: Risk, Compliance, and Accountability in AI-Driven Operations*. PwC, 2024. (CH8)

Qi, Qinglin, and Fei Tao. "Digital Twin–Driven Smart Manufacturing: Connotation, Reference Model, Applications, and Research Issues." *Engineering*, 2023. (CH2)

Rao, Ananya, et al. "Organizational Readiness for Digital Twin Adoption: Skills, Teams, and Change Management." *Technological Forecasting and Social Change*, 2024. (CH9)

Siemens. *Digital Twin Playbook: From Asset Models to Enterprise-Wide Twin-of-Twins*. Siemens AG, 2023. (CH10)

Tao, Fei, et al. "Digital Twin and Industrial Metaverse: New Paradigms for Cyber–Physical Fusion." *Journal of Manufacturing Systems*, 2025. (CH10)

Thoben, Klaus-Dieter, et al. "Lifecycle Management and Federation of Digital Twins in Global Supply Chains." *International Journal of Production Research*, 2024. (CH4)

Verdouw, Cor, et al. "Digital Twins in Logistics and Supply Chain: Real-Time Visibility and Scenario Planning." *Computers in Industry*, 2023. (CH4)

Wang, Shiyong, et al. "Edge–Cloud Architectures for Real-Time Digital Twins in Industrial IoT." *IEEE Internet of Things Journal*, 2023. (CH3)

World Economic Forum. *Digital Twin Cities: Governance, Data, and Public–Private Collaboration*. World Economic Forum, 2024. (CH8)

Xu, Xun, et al. "Data Fabrics and Data Mesh for Digital Twin Platforms in Industry 4.0." *Journal of Industrial Information Integration*, 2024. (CH6)

Yang, Lin, et al. "Model Risk Management for AI-Enabled Digital Twins in Regulated Sectors." *AI and Society*, 2025. (CH7)

Zhang, Rui, et al. "Physics-Informed Neural Networks for Industrial Digital Twin Applications." *Applied Energy*, 2023. (CH7)

Zheng, Peng, et al. "Enterprise Digital Twin: From Single-Asset Models to Organizational Twins." *Computers & Industrial Engineering*, 2024. (CH1)

Zhou, Meng, et al. "Security Architectures for Digital Twin Platforms in Critical Infrastructure." *IEEE Transactions on Industrial Informatics*, 2023. (CH8)

Zhuang, Chen, et al. "Federated Digital Twins: Scaling Across Plants, Regions, and Business Units." *Journal of Manufacturing Systems*, 2025. (CH10)

www.ingramcontent.com/pod-product-compliance
Lightning Source LLC
LaVergne TN
LVHW010607100826
845148LV00014B/2887